Remaking Planning

The Politics of Urban Change

Second edition

Tim Brindley, Yvonne Rydin and Gerry Stoker

London and New York

First published 1996
by Routledge
11 New Fetter Lane, London EC4P 4EE

Simultaneously published in the USA and Canada
by Routledge
29 West 35th Street, New York, NY 10001

Typeset in Bembo by Intype, London

Printed and bound in Great Britain by
TJ Press (Padstow) Ltd, Padstow, Cornwall

British Library Cataloguing in Publication Data

A Catalogue record for this book is available from the British Library

Library of Congress Cataloging in Publication Data

A catalogue record for this book has been requested

ISBN 0–415–09874–2

Contents

Preface

This book arose out of the three authors' parallel teaching and research interests, while based in three separate Schools at Leicester Polytechnic, as we struggled to come to terms with the fate of planning under the reforming Thatcher governments. We found that we had each started to investigate particular examples of planning policy and practice, in an attempt to understand the newly emerging forms of planning. As a team we had the advantage of complementary disciplinary roots in politics, economics and sociology. Together we gradually developed a common perception of the general direction of planning in the 1980s, starting from the recognition that it had fragmented into a number of distinctive and competing styles. Over lunchtime discussions and doodles on table napkins we worked out a preliminary classification. This led to the selection of the case studies, which enabled us to test and develop our ideas and eventually to formulate a typology. As it turned out, as soon as we started to write the book two of us moved away from Leicester and collaboration became rather more difficult!

Each of the authors worked on two case studies, which were therefore mainly the product of individual research, although they developed within the common framework and through a process of mutual criticism. Yvonne Rydin was mainly responsible for Chapter 3, on Cambridge, and Chapter 4, on Colchester; Tim Brindley was principal author of Chapter 5, on Coin Street, and Chapter 6, on London Docklands; and Gerry Stoker was the main author of Chapter 7, on GEAR, and Chapter 8, on Stockbridge Village. All other chapters were written jointly.

The case study method was chosen because it suited the project that we had set ourselves, namely to provide a relatively detailed account of the varieties of planning practice in the Thatcher years. This should be of interest to all practitioners, critics, students and would-be reformers of planning. We hope, too, that each case study stands on its own as a story or portrait of planning in particular local circumstances, and that this will make the book especially useful to students as a complement to more theoretical or abstract planning texts.

Preface to the Second Edition

The aim of the second edition of this book is to review changes in planning policy and practice since the late 1980s, and in particular to see how far our original framework of local case studies and planning styles remains valid. Since 1989 there have been substantial changes in the factors affecting planning. The 'Thatcher years' came to an abrupt close in 1990, so we have dropped the original subtitle of this book – it seems quite inappropriate to refer to the 'Major years' in the 1990s since the present Prime Minister has not dominated British politics to the same extent as his predecessor. Post-Thatcher, however, the British political scene has changed markedly, with a withdrawal from the more extreme ideological positions of both left and right, and the appearance of more common ground between the main parties. This new centrism has been influenced by many developments, including the convergence policies of the European Union after Maastricht, and the search by the Labour Party for a basis for re-election after 17 years of Conservative government. Europe has had other influences on planning policy in the 1990s, notably the rise to prominence of environmental issues and the incorporation of environmental protection policies into UK legislation. Coupled with major economic and social change, and marked shifts in the local politics of urban governance, there is much to take into account in explaining the new trends in planning. We have attempted to discuss the impact on planning of all of these factors in the new edition.

The additional material in the second edition is confined to an extensive Postscript. We have left the original conceptual framework and cases studies unchanged to stand as an analysis of the 1980s. While the conclusions were particular to that time, we still consider that the issues we raised remain valid areas of debate for the future of planning. In the Postscript we have attempted to do two things: first, to review the changing context of planning policy and practice and to identify the concerns and issues which are driving the planning agenda in the 1990s. The dominant influences remain those of the economic and political contexts, which we stressed in our original typology of planning styles,

PREFACE

but we have added further comment on the social and environmental contexts as important additional influences today. Second, we have returned to our six styles to see how planning has responded to its changed context. Our main argument is that the diversity of planning styles in the 1980s has been reduced to two dominant styles in the 1990s. In bringing the picture up to date, we have reviewed the general reforms in the planning system, new legislation and changes in national policy. We have also revisited and analysed those case studies which best illustrate the changed approaches at the local level.

The updating of the case studies deals in particular with the London Docklands and Coin Street. In the 1980s, these cases were at opposite extremes of our continuum of planning ideology, representing market-led leverage planning and the market-critical style of popular planning respectively. In the 1990s, they still show distinctively different approaches to urban regeneration but they also illustrate how planning styles have tended to converge towards a more central ground of planning practice, pursuing the implementation of policies through a variety of partnerships.

Readers familiar with the first edition should turn to the Postscript for our analysis of the 1990s and new case study material. Readers coming to this book for the first time will see that it is in two parts, Chapters 1–10 dealing with planning in the 1980s, and the Postscript addressing the issues of the 1990s.

We wish to acknowledge our debt to information and comments from the following individuals in preparing the second edition: Bob Colenutt, Robert Cowan, David Henshaw, Ben Kochan and Iain Tuckett. We are also grateful to our editor, Tristan Palmer, whose consistent encouragement has helped to advance our argument and bring this new edition into being.

Tim Brindley, De Montfort University
Yvonne Rydin, London School of Economics
Gerry Stoker, University of Strathclyde

Acknowledgements

We have benefited from the help and support of many people in writing this book. First, we must acknowledge the time and resources made available within our respective institutions, Leicester Polytechnic, North East London Polytechnic and the Institute of Local Government Studies, University of Birmingham. We would also like to extend our personal thanks to those people who have been especially helpful in providing information, opinion and advice on particular topics and on the major themes of the book, here listed in alphabetical order:

Douglas Aikenhead, Robin Boyle, Andrew Church, David Clapham, Michael Hebbert, Keith Kintrea, John McHale, Sarah Monk, Peter Mottershead, George Nicholson, Peter Thornton, Louanne Tranchell, Iain Tuckett and Stephen Young.

We have also benefited from the opportunity to present our developing ideas at various seminars and conferences, and we are grateful for the comments and encouragement of those who responded.

We would like, as well, to acknowledge the contribution of employees and representatives of the following organizations who provided valuable assistance and information for our research:

Cambridge City Council, Cambridge Urban Studies Centre, Colchester District Council, Colchester Urban Studies Centre, Coin Street Community Builders, Docklands Forum, Glasgow District Council, Knowsley Metropolitan District Council, London Docklands Development Corporation, Scottish Development Agency, South Cambridgeshire District Council and Stockbridge Village Trust.

Thank you to Alan Sankey for drawing the maps.

And lastly, a very big thank you to Lynne, George, Deborah, Bethany and Robert.

It goes without saying that the three authors alone are responsible for the interpretations, arguments and errors which follow.

1

Introduction

This is a book about planning in Britain in the 1980s, something which many think no longer exists. It is said that the Thatcher governments have all but abolished planning since 1979. The relaxation of many controls, the introduction of enterprise zones and simplified planning zones, the transfer of planning powers to urban development corporations, the greater stress on market criteria in development control decisions, all have been taken as lethal attacks on planning itself. Ravetz, for example, observes that:

> the Thatcher Administration...is fast dismantling much of the planning system, along with many other parts of the Welfare State. This puts planning on trial, so to speak, for its life. (1986, p. 9)

For Ambrose (1986), who asks *Whatever happened to planning?*, the execution has already been carried out and it is time to write the obituaries.

Yet if we look at what has actually happened to planning in the past decade, we find that reports of its death are greatly exaggerated. Planning is still being practised, there has been no major reform of the Town and Country Planning Acts of 1968 and 1971, and development plans still have a significant role. Most of the changes in planning have been either revisions of policy within the existing system, or additions to the system, often involving both state intervention and public expenditure. While there has been a sustained attack on planning from the New Right, this has been vigorous in its rhetoric but rather less drastic in its actions. Planning has certainly changed, but it has not yet been eliminated.

The 'death of planning' thesis reflects a certain loss of perspective. Commentators from both centre and left political positions, and from within the planning profession, seem to have become imbued with a romantic notion of planning as if it were a uniquely social democratic, or even socialist, idea. They tend to look on planning legislation and policies as essentially idealistic and progressive, favouring the poor and powerless over the rich and powerful.

1

Where planned development has fallen short of these ideals, and it often has, this is put down to a 'failure' of planning rather than an intended outcome. Proponents of this thesis appear temporarily to have forgotten that land-use planning has pursued goals of economic efficiency and maximizing land values as much, if not more than, those of social justice and equality. Consequently, they have been unable to recognize the policies of the New Right as 'planning' at all. This leads them to ponder the 'paradox' of a government which simultaneously criticizes planning and creates highly interventionist bodies such as the urban development corporations. But it is our contention that the paradox is a false one and that, despite the rhetoric, the Thatcher government is not anti-planning in the broad sense. Its attack has been on collectivist, 'welfare' or, as we term them, market-critical conceptions of planning, and its demands have been for new forms of planning more oriented to the market and the interests of developers.

It is the central theme of this book that the 1980s have witnessed first the fragmentation and then the remaking of planning, which is emerging from the past decade with its goals and purposes reorientated. Our argument springs from a broad definition of planning, which we take to refer to all activities of the state which are aimed at influencing and directing the development of land and buildings. In this sense, state intervention can be concerned with many different purposes, managed through diverse institutions, and can bring into play a variety of social and economic interests. The policy processes associated with state intervention are complex and conflict-laden, but they are always central to the direction of urban development and renewal. Our book is therefore concerned with the politics of urban change, focusing on the struggle between different forms of state intervention and the restructuring of planning styles that has taken place in the 1980s.

The change in the direction of planning has not happened cleanly or swiftly. While the central government has attempted to change the framework of planning policy and legislation, within this framework local authorities and local communities have continued to pursue their own, often quite different, goals. The result has been a rather confused picture, with a wide variety of approaches to planning being pursued simultaneously in different areas, and sometimes competing for dominance in the same area. Before we try to bring some order to this confusion, we need to consider how such a major change in the direction of planning came about. It emerges that it was not simply the result of the rise to power of a government committed to a particular ideology, but that it was rooted in a 'crisis' in planning and the context in which it was operating in the 1970s.

The crisis in planning

We can begin our examination of planning's crisis with the inside view, looking at how it affected the profession. Ravetz (1980), in a perceptive and wide-ranging history of postwar planning in Britain, gives an account of the state of the profession in the mid-1970s. By then it was hard to find anyone with a good word to say about planning, and the profession was growing increasingly demoralized. She cites in particular the Town and Country Planning Association's 1977 report, *The crisis in planning* (Ash 1977), which recorded a 'public disillusionment with planning so widespread that one does not even feel obliged to document it'. Planning had failed to live up to its own claims and nobody's expectations seemed to have been satisfied. For some, it had failed to achieve the wholesale modernization of the built environment that it had so enthusiastically championed since the 1940s. The 'evangelistic bureaucrats' (Davies 1972) had run out of steam and the country was littered with half-completed urban motorways, unfinished slum clearance projects and partially redeveloped city centres. For others, it was the failure of planning to prevent undesirable development that was its chief weakness, whether in the form of surplus office blocks, industrialized council housing or the destruction of historic buildings. The whole direction of planning was being challenged, by grass-roots community activists and middle-class conservation societies. However much or little of this could be blamed on the statutory planning system, professional planners bore the brunt of the criticism and faced repeated accusations of failure. No wonder the profession began to lose confidence.

In the face of this unrelenting criticism, planners found it hard to know where to turn for their defence. As Ravetz points out, the real weaknesses of planning as a profession were revealed in its exaggerated claims to knowledge and expertise, and in its subordination to direct political control within state bureaucracies. In their claim for professional status, planners had pretended a greater knowledge of the processes of land and property development, and therefore of 'the future', than available disciplines could provide. They had also gone down the self-deluding path of 'affecting a fastidious political neutrality' (Blowers 1986, p. 16). As well as being exposed as a sham by academic criticism and political opposition, this had also left planners vulnerable to political manipulation by powerful interests. Too often they had appeared as charlatans in the pockets of the property development industry. The truth is that planning as a profession had become too closely associated with one set of goals, one approach to the future of the built environment,

3

an ideology which Ravetz characterizes as the 'clean sweep' style. Planning had come to stand for wholesale change, but in the 1970s it began to emerge that not everyone wanted change on this scale, and that economic circumstances were going to make it much more difficult to achieve.

Opposition to change was opposition to the planners' vision of modernization. In the 1960s the planning profession had taken up the banner of modernization in an evangelistic spirit. Davies (1972) has pointed out how successive leaders of the profession, such as Colin Buchanan and Wilfred Burns, proclaimed an image of planning as the means to a better future. It was the duty of the planner to convince doubting fellow citizens to let go of the past and welcome the future, in all its concrete reality. As the voices of the objectors grew louder, the planning system offered 'participation' as the means to strengthen the consensus behind planning proposals. Instead, participation and protest demonstrated the blatant lack of consensus for change and exposed the political biases of the planners. The protest groups were varied and represented a wide range of interests. Some of the most vociferous were middle-class property owners objecting to motorway routes, but working-class residents also objected to the destruction of their inner-city neighbourhoods for speculative office development. Both professionals and tenants decried high-rise industrialized council housing. Local campaigners opposed the unnecessary destruction of established communities in slum clearance programmes. Middle-class and upper-class supporters of the burgeoning conservation movement helped to save areas such as Covent Garden and Bath from further destruction.

The lack of consensus for change put a major brake on development in the 1970s. Plans for redevelopment were suspended or reversed, slum clearance was replaced by gradual renewal, and the spirit of modernization suffered a major setback. If, as Cullingworth (1985) has argued, the new development plans system of the 1968 Act depended on consensus, then the absence of that consensus left its products – structure and local plans – indeterminate and vague. Where there were strong demands for change it was resisted, and where there was a need for change, it was compromised or neglected.

It was not only the expression of public attitudes which altered the pace and scale of development in the 1970s; it was also a major change in economic circumstances. If the modernization of the built environment, in the forms offered by the planners, was not universally desired, then neither was it any longer achievable. This began to become clear after the financial crises of the late 1960s when the long period of postwar economic growth first seriously faltered. The oil crisis of 1973 and the ensuing recession killed off most remaining plans for large-scale development and urban renewal. This exposed another underlying weakness of the planning system, its dependence on economic growth. The development plans system was based essentially on state regulation of private sector development. Where the state undertook development, this was mainly the provision of physical and social infrastructure (roads, schools, hospitals and housing), or else it was in partnership with the private sector. Consequently, when economic crisis pushed the private sector into recession and indirectly produced a major retrenchment in the state's direct role in development, there was little left to plan for.

Planning might have recovered from a temporary setback in growth in something like its old form, even overcoming much of the public resistance to change. But the 1970s quickly turned into a period of deep and prolonged economic decline, and this was something for which planning had few remedies. The gradual decline of northern England, Wales and Scotland had been apparent for some time. Successive governments had used regional aid and state development projects in an attempt to stem this decline, but to little effect. The 1970s saw a rapid rise in the rate of decline, particularly of manufacturing industry, coupled with the recognition that it was seriously affecting all of Britain's old industrial cities (Lawless 1981).

The processes of economic and industrial restructuring had a dramatic effect on particular localities, enormously increasing the disparities between different places. While some cities and towns experienced growth and new patterns of employment, others experienced massive decline and very high levels of unemployment. Areas such as the West Midlands suffered from the collapse of key sectors of manufacturing, including the machine tool, engineering and car industries. During the 1970s the industrial base of Birmingham shrank by a third (Spencer *et al.* 1986). In Sheffield in 1971 there were 139 000 people employed in manufacturing industry. Ten years later the number had declined to 90 000; and by 1987 it had collapsed to 58 000 (Sheffield City Council 1987, p. 7). In contrast to these areas of decline, Boddy *et al.* (1986) describe the experience of Bristol and the surrounding M4 growth corridor. Here, while traditional manufacturing declined, there was a considerable expansion of service sector industries, combined with the rise of newer activities based on electronics and high technology, leading to claims of an economic renaissance. One effect of these changes, which became manifest in the 1970s and continued into the 1980s, was to make 'locality' more significant (cf. Massey 1984). They brought different problems to the fore in different locations, lending

5

support to the rise of new styles of state intervention and planning to meet these diverse challenges.

By the end of the 1970s the crisis in planning was deeply rooted and comprehensive in its scope. The two main supports of planning as an enterprise, a broad consensus in favour of change and economic growth to generate change, had both been seriously undermined. Planning was left exposed, vulnerable and confused; but it could not be abandoned. Whatever interests were in control of the various parts of the state, they would demand some sort of planning to ensure that their version of the future prevailed. A search began for new forms and styles of planning, to meet the needs of different localities, to bring about patterns of development desired by various interests and to match the political rhetoric of those interests. It was this process of evolution and experimentation which gave rise to the varied styles of planning which have characterized the 1980s, and of which this book attempts to give an account.

Structure of the book

The argument of this book is that, in response to changed social and economic circumstances, planning fragmented during the 1980s into a range of different forms. Chapter 2 provides a six-fold classification of the planning styles of the decade. Chapters 3–8 present detailed case studies of each of the styles in practice. The case studies show the main features of each planning style, focusing in particular on the institutional arrangements, types of politics, and conflicts and tensions associated with the different forms of state intervention. Chapter 9 sets out to compare the different planning styles in the light of the evidence presented in the case studies, developing the discussion of their effectiveness and outcomes. Chapter 10 concludes by arguing that the election in 1987 of the third Thatcher government confirms the new direction of planning, which is being remade with a predominance of market-oriented styles. We examine the nature of planning as it moves into the 1990s, the likely impact of the new approach, and how an alternative agenda might be established.

2

The fragmentation of planning

We have argued that the changed circumstances of the late 1960s and early 1970s led to a crisis in planning, a collapse of confidence amongst both the public and professionals. This crisis is now manifested in the fragmentation of planning into a number of distinct approaches. It is our contention that the 1980s mark a turning point in the postwar history of planning. Previously, planning had been diversified in practice, with different local authorities developing their own policy variants and with localized experimentation. However, this diversification occurred within the context of a unified debate about planning, a debate which focused on the development plan system and the decision-making practices of professional planners. There was a general consensus on the role of the planning system, in terms of broad goals and means. Arguments concerned relatively minor procedural matters or rarefied planning theory.

The past decade has seen a heightening of economic and political conflicts within society, and this has been reflected in planning. The debate over planning has splintered as the lines of current economic and ideological cleavages have become more sharply delineated. A variety of new and old approaches to planning now vie with one another. This represents a moment of transition in planning history as one dominant ideology of planning attempts to replace another. In the meantime it is sometimes difficult to see anything other than a confusion of competing ideas, each promoted by a sectional interest. The purpose of this chapter is to inject some clarity into the current confused state of the debate, to identify the competing approaches and relate them to the prevailing economic and ideological cleavages.

We identify six styles of planning. Each style represents a particular stance in the debate on planning and proposes a particular mix of policy goals, working methods and identity for the planner. Some styles are strongly influenced by a radical vision and have the character of blueprints for local experiments. Other styles are not

7

so new but rather derive from adaptations and modifications of established planning methods. Our central argument is that these various styles together capture the essence of the current state of planning, albeit in a simplified form.

It is important to recognize the limits of this planning debate. By and large, it has not focused on radical alternatives to the present system whether from the Right (Sorenson 1983) or the Left (Ambrose & Colenutt 1975). It accepts the liberal democratic framework of an interventionist state existing alongside a reliance on market operations, and puts forward proposals for dealing with the resulting tensions. In doing so each approach recognizes, at some level, the inevitable interrelation between the state and the market, that the market requires the support of state policies, and that the state relies on the market to produce many policy outcomes. Certain approaches are closer to a radical alternative than others and may even appear to disguise this underlying tension. But, as the case studies reveal, the tension surfaces at the implementation stage even if the rhetoric seeks to avoid it. In clarifying the planning debate of the current decade, we are therefore concerned to chart the plurality of proposals within the political framework of a liberal democracy. Other commentators have noted 'the increasing apparent variety in planning practice' (Healey 1983, p. 271) and the competition between a number of different proposals for the planning system (Nuffield Commission of Inquiry 1986, Ch. 7). One way to organize this plurality is through the use of a typology. Our typology relates the six styles to the prevailing economic and ideological cleavages. These dimensions and the characteristics of the six styles are set out in the rest of this chapter.

A typology of planning styles

The typology is set out in Table 2.1. It is developed along two dimensions, which represent contemporary ideological and economic cleavages. The first dimension reflects the break-up of the ideological consensus on which postwar planning has rested. As the Nuffield Commission noted, there is still within planning circles a great desire to assume a consensus (Nuffield Commission of Inquiry 1986, p. 97) but the reality of planning debate is that there is a sharp distinction between proponents based on their attitude to market processes:

> ...we have to distinguish between planning that takes a positive view of the market, while attempting to correct inefficiencies, and planning that

takes a positive role in attempting to redress the inequalities of the market and to make good its omissions by measures to increase the access of the disadvantaged to housing, health, recreation and communal activity. This is one of the most important of the dimensions of disagreement which we shall analyse...(ibid., p. 184)

Table 2.1 A typology of planning styles

Perceived nature of urban problems	Attitude to market processes	
	Market-critical: redressing imbalances and inequalities created by the market	Market-led: correcting inefficiencies while supporting market processes
Buoyant area: minor problems and buoyant market	regulative planning	trend planning
Marginal area: pockets of urban problems and potential market interest	popular planning	leverage planning
Derelict area: comprehensive urban problems and depressed market	public-investment planning	private-management planning

In styles which embody a positive view of the market, demand as measured by the consumer's purse is the main indicator of where and when development should occur. The market mechanism determines who receives what and at what cost. The main actors are in the private sector and profit is the motivation for their actions. Such market actions require a framework of state support, and planning policies are one way of providing this. Occasionally, where market outcomes are judged to be inefficient, additional planning powers will be brought into play. Nevertheless, market processes are considered to be a satisfactory mechanism of allocation in the majority of cases, indeed vastly superior to alternative mechanisms.

By contrast, in styles critical of the market, the outcomes of market processes are considered to be partly or even wholly unacceptable. The inequalities resulting from such processes are stressed, creating a need for planning policies to redress them. Planning is also needed to rectify imbalances, such as that between short- and long-term perspectives on resource use. This requires an organization which is not simply responding to market indicators but which will take a dominant role in defining the needs to be met and even in meeting those needs. The market mechanism may therefore be replaced in the pursuit of a more generally defined goal of welfare.

The second dimension of our typology reflects the impact of economic change over the past decade or more. As noted in Chapter 1, economic recession and the associated restructuring have had an uneven spatial effect and created increasingly sharp divisions between regions and localities. This is associated with a varying level of private sector interest in land and property between areas. The planning debate has sought to identify what should be the appropriate solutions for different areas facing different problems. Our categorization follows that of the Property Advisory Group report, *The structure and activity of the property development industry* (1980).

First, there are areas where the industry will invest without any public-sector support or subsidy. These might be termed 'buoyant' markets, of which the prime examples currently are sites for suburban housing schemes in the South East and large-scale out-of-town retail development in almost any part of the country. Secondly, there are areas where the industry could be induced to invest with appropriate support and subsidy from the public sector. These we can term 'marginal' areas in which the development industry has less confidence. Often these are areas which have suffered from prolonged periods of neglect but where the immediately surrounding area is such that the potential spread of economic activity may rekindle private-sector interest. Thirdly, there are areas where no subsidies can induce the development industry to invest. These 'derelict' areas have been abandoned by the private sector and are now viewed as 'no-go' areas. The Property Advisory Group described them as 'areas which are either unattractive or where there is little prospect of them becoming attractive' (para. 7.28).

Recognition of the particular impact of economic restructuring in different areas, together with the ideological split between market-led and market-critical approaches, characterizes the fragmentation of planning into distinct styles. So, in more prosperous areas, planning of a negative kind is considered most appropriate. Development interests consider the problems found in such localities to be minor ones, a view that is commonly shared by local middle-class residents. This does not preclude the existence of severe problems in terms of housing and employment opportunities for some of the local working-class population. However, these issues frequently do not find a political voice, and local advantages are seen to outweigh disadvantages. Planning is therefore largely directed towards improving or preserving existing living conditions.

The private sector exhibits considerable investment interest in this type of locality, and planning styles are consequently concerned with reacting to private development initiatives, not with actively encouraging them. Control and regulation are the key planning tools. This is true of both regulative and trend planning, the difference between the two styles being primarily one of degree. Regulative planning involves an attempt to control and direct market pressures in order to manage urban change in the public interest. Trend planning, by contrast, does not try to redirect market forces but applies minimal planning powers to facilitate development in line with market pressures.

In marginal areas a greater degree of positive planning is needed to bring the problem area up to the economic standards of the surrounding area. Sites of potential interest to the market may be satisfactorily developed but planning has to stimulate the change. Debate focuses on the planning mechanisms by which the public and private sectors are brought together to undertake this task of restructuring the local market. In popular planning the public sector is dominant, but acting primarily through the community rather than through government institutions. In leverage planning, the public sector also plays a significant role but the private sector is seen as the main agency of change. The two styles may be contrasted as, on the one hand, an attempt to regenerate community and public sector interest in the development of an area; and on the other hand, an attempt to regenerate an active private market, essentially by altering market conditions to make investment less risky and more profitable.

In derelict areas there is a widely perceived need for large-scale action to reverse, or at least manage, the urban decay, so there is a shift towards a totally planned local environment. Rather than the atomistic decision-making typical of market processes, the derelict area is brought under the control of one agency. Public-investment planning, in which the public sector redevelops an area by purchasing land and providing all or most of the capital investment, is favoured by some groups on the political Left. Private-management planning, on the other hand, involves private-sector agencies taking control of an area, even where public-sector assets are involved, and is a style favoured by the Right.

More than any of the other 'pairs' of planning styles, public-investment planning and private-management planning are in direct competition. Both are presented by their advocates as solutions to the worst problems that planning has to face. The first represents the claim that only the public sector can deal with these problems, the second the claim that only the private sector can provide the solutions. Given the definition of derelict areas offered by the Property Advisory Group, private-management planning in particular represents an act of unusual faith in the private sector.

The scope of the typology

In presenting a typology, our aim is to clarify the current planning debate, to organize the various proposals in relation to each other and to establish their essential characteristics. While we would argue that such a typology has certain advantages, it is nevertheless the case that any typology is limited by its own simplifying assumptions. We must emphasize that we are not trying to capture the complexity of planning thought and practice in a simple matrix. The complexity of the social processes involved in both practice and thought and of the relationship between them has to be recognized. There are three particular limitations which we want to acknowledge at the outset.

First, while the typology has an economic dimension, it does not provide a full representation of the processes which structure the local economy. These economic processes operate at the national, regional and local level to produce specific local economies and urban environments. Each locality has its own history of the interaction of these processes, distinct from any other locality (Rees & Lambert 1985). Furthermore, space is actively used in location strategies by the agents of urban change, as Massey (1984) has shown. The complexities of these multilevelled processes cannot be captured in three area-specific categories. Rather, our typology identifies the dominant perception of a particular area's local economy. We are not suggesting that by identifying that perception it is possible to read off in simple terms an appropriate planning style, for any local economy will contain sites where market demand is strong and other locations where no development interest exists at all.

Secondly, the typology presents a simplification of ideological stances as they operate at the local level. A distinction is proposed between approaches which are supportive of market processes and those which are critical. Although the Left may be associated with market-critical styles and the Right with market-led ones, it is not possible to reduce political allegiance to this one dimension. Again, the involvement of public sector agencies or public sector resources does not imply a market-critical approach, nor the involvement of the private sector a market-led approach. In practice there is a wide variety of ways in which the state supports, restructures and replaces the market. It is not always easy to identify the dominant ideology in any locality or to tie it down to one side of a market-led/market-critical divide.

Thirdly, the typology makes no claims as to explanation of the relationship between policy debate and policy practice. Many factors will influence the operation of planning in a particular locality. It is not intended that a particular form of planning practice in a locality be directly deduced from the two dimensions of the typology. Rather, the typology helps to organize ideas about planning. Planning practice is explored in the substantive case studies.

Despite these caveats, the adoption of such a typology does have distinct advantages. First, it simplifies the complexities of ideological debates and economic processes acting over space to a limited and manageable set of categories. Secondly, these categories reflect the terms in which public debate on planning is conducted, a debate which is often oversimplified in its understanding of urban processes. For example, British land policy has tended to operate in terms of planning for spatially delineated areas as in General Improvement Areas, Enterprise Zones or local plan areas. It can be argued that this tendency directs attention to local features of the area as the cause of urban problems and diverts attention from the broader economic processes (Parsons 1986). Nevertheless, planning thought and practice remains largely caught within the constraints of area-centred policies.

Thirdly, the typology provides a basis for identifying case studies for further analysis. It is in these case studies of planning practice that the complexities of economic and political processes can be explored. The ways in which the actual styles of planning evolve through implementation can be investigated (Barrett & Fudge 1981). In the case studies it will become apparent that two or more of the styles identified in our typology may be jointly operated in practice, that practice can move along a continuum between styles, and that no one style is uniquely identified with any one locality for any length of time. We do, however, suggest that the typology gives useful guidelines as to the dominant competing styles likely to be found in particular local economic circumstances during the 1980s. Before moving on to the case studies, we examine each of the styles in more detail.

Regulative planning: adapting to changed circumstances

Regulative planning lies at the heart of the postwar planning system established in Britain in 1947. Under planning legislation, local authorities have two principal functions. First, they are required to draw up plans for future development and land use. Second, they are given power to grant or refuse planning permission for most private-sector building and redevelopment schemes. It was argued that, by using these twin powers, local authorities could guide urban change so that it fitted their planning blueprints. To some degree British planners have been successful in this aim, particularly where they have sought to restrain development and contain urban sprawl as in Green Belt policy (Hall *et al.* 1973, Best 1981, Munton 1983). By reacting to private-sector initiatives many local authorities have exercised considerable negative control and displayed some potential for redirecting demand in line with public plans, in partial opposition to market forces.

This regulative planning style is central to the ideology of the planning profession. It enables the planner to pose as the expert manager of the urban system, a role buttressed by a number of ideological components, principally the rational and systematic mode of decision-making adopted by planners. This role also involves the assumption of an underlying consensus within society so that, in the face of competing interests, planners can claim to reach a judgment in the best interests of all (Simmie 1974, Ravetz 1980).

However, rational regulative planning in the public interest has not gone unchallenged. Criticisms have emerged from a variety of sources (some of which are taken up in the next section on trend planning). Left-wing analysts have emphasized the weakness of many aspects of the development control system, arguing that such planning follows rather than directs the market. Certainly, regulative planning is most effective when local demand is buoyant, since it can do little to stimulate private-sector initiatives. Some argue that the apparent postwar successes of planning control have in fact been related to the pattern of public investment in infrastructure, particularly in the New Towns (Ball 1983).

Even where development control is found to be affecting private-sector decisions, it has been argued that the outcomes are very unevenly distributed. Landowners, developers and owner–occupiers appear to be benefitting at the expense of others, such as the homeless, council tenants and the unemployed (Simmie 1981). The claim of planners to be acting in the public interest looks more suspect in the context of a society where conflicts of interest and extensive inequalities persist (Simmie 1974).

From the viewpoint of property developers and the political Right, regulative pl nning has been criticized as unnecessarily restrictive and constraining for the private sector (DoE 1975, House of Commons 1977). This criticism has been reinforced by an intellectual attack on the ability of the state to intervene rationally and effectively. Instead, New Right economists have re-emphasized the primacy and efficiency of the market (Adam Smith Institute 1983, Sorenson 1983, Green 1986).

Despite these criticisms, regulative planning remains the dominant image of planning, and the tools of the system are essentially geared to this end. Effective regulative planning is based on hierarchical strategic planning and a range of development control powers. While planners no longer expect to have total control over the pattern of urban change, they still seek to control individual private-sector developments in pursuit of public policy goals. However, the excesses of 'scientific' decision-making and the comprehensive approach have been downgraded. There is a new emphasis on negotiation and network-building skills in planning education (Underwood 1980, 1981). Through these skills planners seek to influence development proposals before and after planning applications are received, and to extract community benefit in the form of planning gain.

Trend planning: streamlining the system

The expression 'trend planning' was first coined in the 1970s by analysts keen to emphasize the powerlessness of regulative planning. In the aftermath of the property boom, development control was seen to have retreated from a directive, if reactive, system to a passive and completely ineffectual one (Broadbent 1977, Pickvance 1981). The tools available were seen as both clumsy and weak (Ambrose & Colenutt 1975, Kirk 1980). Planners were frequently subordinate in their dealings with property companies and developers, being easily persuaded and led (Dumbleton 1976, Wates 1976, Goldsmith 1980). In particular, the lack of public control over both land and investment funds meant that development control could not live up to its name. As Pickvance commented:

> ...in city centre business and financial districts most planning authorities would not consider any other sort of development besides offices. In other words, certain types of land use are seen as 'logical', 'sensible' and 'financially sound'. In city centres it is seen as 'illogical' to zone land for uses which are not the most profitable and which do not bring in the highest rates incomes. (1981, p. 70)

In short, in many development plans the allocation of land has been very similar to that which would have occurred under market forces.

Trend planning now describes a head-on challenge to the existing regulative style, attempting to reorientate it to a private-sector perspective. In this form of planning, the negative powers of planners are not used to restrain or to bargain. Rather, development plans consciously reflect market trends in the allocation of resources, and planners are charged with facilitating development in line with market demand.

This style of market-led planning has been strongly promoted by the Thatcher administrations since 1979. It emerges in the concern to streamline the planning system and reduce delays (Thornley 1981), in the debate about the release of land for private housebuilding in Green Belt locations and areas of high demand (House of Commons 1984, Rydin 1986), and in the explicit introduction of market criteria into development control decisions (DoE 1980). The priority is private-sector development activity and responsiveness to market forces. As one recent Conservative Secretary of State for the Environment has commented:

> Planners must help create the right conditions and ensure that consumer initiatives prosper ... planning procedures should not hamper the economic recovery ... Planning authorities must adopt a flexible and pragmatic approach to meet the ends of versatile enterprises ... I am determined that all planning authorities should be sympathetic to ... industry. (Jenkin 1984, pp. 15–16)

Where necessary, local discretion has been reduced to enforce the adaptation of the planning system to this style, through structure plan modifications, planning appeal decisions and the call-in powers of the Secretary of State for the Environment.

The most recent attempt to impose trend planning is contained in the proposals for Simplified Planning Zones (SPZs) in the 1986 Housing and Planning Act. Under these proposals, SPZs can be identified by local authorities or by the Secretary of State using default powers. A scheme of permitted uses will be prepared for each zone and any conforming development will not require planning permission. The process of development control will be completely bureaucratized. However, as our case study will show, there are political pressures for maintaining a degree of environmental control which currently hinder attempts fully to streamline planning decisions. The schemes for SPZs will therefore undergo procedures very similar to those for adopting local plans, including a public local inquiry where there are objections.

The likely impact of SPZs on the built environment is generally agreed to be minimal. The Royal Institution of Chartered Surveyors regard the proposals as a missed opportunity. The Royal Town Planning Institute has not been unduly worried, although it has put forward its own ideas for off-the-peg planning permissions. Other commentators agree that SPZs are not a radical departure in planning practice (Thornley 1986). Robinson & Lloyd conclude:

> The ideological significance to the government of SPZs would seem likely to be greater than their practical impact where the concept is adopted. (1986, p. 63)

It can be argued that these limits to the SPZ proposals reflect the role that planning can play in supporting the market. Trend planning in structure and local plans helps private investors and developers to coordinate and manage their investment plans (Farnell 1983). Thus land-use zoning and development control can be regarded as to some degree supportive of the operation of land markets by maintaining land and property values in some locations and preventing anarchic and damaging competition.

Trend planning therefore currently represents the end result of reorienting regulative planning from the public interest to the private interest. The response to the exposure of weaknesses in development control is not to reform or strengthen it, but to strip it to the bare bones. Only those aspects of planning are retained which seem to be functional for private development or which are electorally sensitive. As such, trend planning is only suited to areas broadly free of urban problems.

Popular planning: reviving the community

Popular planning is rooted in the public challenge to major planning proposals which emerged during the late 1960s. This took the form of organized opposition to development which threatened local communities, including slum clearance (Lambert *et al.* 1978), urban motorways (Hart 1976) and large-scale commercial developments (Wates 1976). This protest produced a large number of local action groups and campaigns, each fighting a specific issue. The Campaign for Homes in Central London (1986) identified 11 neighbourhood groups formed around housing issues alone between 1970 and 1974. At the same time the 1968 Town and Country Planning Act introduced statutory public participation in planning. Techniques for publicizing plans and consulting the public were recommended in the Skeffington report (Ministry of Housing and Local Government 1969) and adopted by many local authorities. While it may have been intended to defuse popular protest, public participation brought planning issues before a wider audience and provided more opportunities for opposition to be voiced. Public inquiries became the object of demonstrations and disruptive campaigns.

Popular planning seeks to go beyond the defensive anti-development campaign, and even beyond the enhanced consultation and participation procedures of the Skeffington report. Rather, it seeks the formal recognition and eventually the implementation of plans prepared by the local community. To achieve this aim there appear to be two prerequisites: an area of marginal concern to the development industry and a sympathetic public sector agency.

Marginal areas are most appropriate because they provide an economic space within which the community's demands can be satisfied. In more prosperous areas, the strength of development pressure is too great for the community to stand any chance in the competition with developers for sites. This is seen in Christensen's (1979) account of Covent Garden, where a popular plan, which began in opposition to major commercial and public-sector redevelopment, only gained ascendancy after 1973 when the property boom collapsed. Although this plan was eventually adopted as the statutory local plan, it is questionable whether that would have occurred in more buoyant market conditions. Indeed, subsequent growth of market interest in Covent Garden has meant that little of the popular plan has been implemented. So great is the contrast with the present that it is difficult now to recall the marginal nature of the area in the mid-1970s.

The Covent Garden plan also rested on the support of the Greater London Council (GLC) under a sympathetic Labour administration which lost office in 1977. The popular planning movement in general was given substantial encouragement by the election of a radical Labour GLC in 1980. It declared a Community Areas Policy (GLC 1985a) and created a Popular Planning Unit. The GLC defined 'Popular Planning' as

> ...planning from below – planning that is based on people coming together in their workplace and community organisations to formulate their own demands and visions for the future. Popular planning starts with resistance ... The second stage is the formulation of alternatives and the fight to put them into practice. (GLC n.d.)

The GLC also mooted the idea of Planning Action Zones where plans could be jointly prepared by the authority and local communities (GLC 1985b). The purpose of all these initiatives was to explore methods of positive planning for local needs in the blighted or decaying working-class neighbourhoods of Central and Inner London. The policy was based on the assumption that local authorities alone cannot overcome the deficiencies of the market without creating an active role for those people directly affected by the processes of change. Popular planning therefore combines state intervention with an active popular base.

The advocates of this style of planning see many benefits of popular participation. First, it can help to restore confidence to the people of areas subject to declining or fluctuating private-sector investment interest. To quote *The people's plan for the Royal Docks*:

> We have brought together and into the public view, the demands of those who have suffered from false promises in the past. We have done this in order to give people in our area confidence in their own ideas, confidence that they have a right to have a say. (Newham Docklands Forum 1983, p. 5)

Secondly, a plan drawn up after extensive popular consultation is more likely to be in line with local needs. It stands some chance of avoiding the unpopular blunders of the recent past, such as multistorey housing for families and deck-access apartment blocks, and holds out the promise that the knowledge and expertise of local people can be incorporated into decision-making processes.

Thirdly, popular planning can establish a base of political support and pressure which is needed if the planning proposals are to come to fruition. As Wainwright notes, 'support for popular planning has meant helping people develop the confidence and organisational strength to challenge the power of those at the top' (1985, p. 7).

Before the GLC was abolished in 1986, none of its popular planning policies was accepted by the government, which continued to recognize only the 1976 Greater London Development Plan. However, the demise of the GLC does not mean the end of popular planning. The 'New Urban Left' of Ken Livingstone's administration is also evident in various London and other Labour-controlled councils, as well as within certain community groups. The aim pursued by these groups, through popular planning and other policies, is to restructure local economies and neighbourhoods. The tendency in the 1980s has been for local campaigns to form broader alliances with these local Labour parties and with trades unions, pursuing their specific aims under the banner of 'local socialism'(Lowe 1986). This is partly in reaction to the Thatcher government's centralizing tendencies; and partly an expression of a new concept of municipal socialism which is evolving independently on the Left.

Popular planning is not a uniquely socialist ideal. Indeed, its most implacable opponents are found on the 'hard left' (McDonald 1986). Rather, it can be seen as part of a politically diverse movement for neighbourhood revival and local control of resources. Closely related are the community architecture and the community technical aid movements, which aim to bring control over both design and building to the end users of development (Wates & Knevitt 1987). These movements received a boost in 1987 with the election of Rod Hackney to the presidency of the Royal Institute of British Architects. Hackney, himself a pioneer of community architecture, not only championed the cause but also brought it the patronage of the Prince of Wales and the apparent endorsement of the government and business interests. Popular involvement in planning and development has come to represent a moral ground which appeals to most political interests and ideologies. This unlikely consensus at least suggests that there will be continuing opportunities for popular planning in marginal areas.

Leverage planning: stimulating the market

The essential ingredient of leverage planning is the use of public-sector finance to stimulate a weak market and to release a greater volume of private-sector investment. Although the idea of bringing in the private sector has been strongly promoted by the Conservative governments since 1979 (DoE 1982, Heseltine 1983), it is not a new approach. There have been many examples of partnerships between the public and the private sectors, throughout the postwar period. The practice of the public sector clearing sites and providing physical infrastructure to support private-sector investment also has a long postwar history. In various ways the state has been willing to underpin the private sector, effectively subsidizing development schemes that might not otherwise have gone ahead.

However, since 1979 leverage planning has had a more prominent role, particularly within carefully delineated spatial boundaries which define a market capable of stimulation. This was first seen in the establishment of Enterprise Zones and, more recently, Freeports (Massey 1982, Hall 1983, Lloyd 1984). The designation of an Enterprise Zone brought exemption from rates, development land tax (now abolished) and industrial training levies, both for existing land users and businesses and for those wishing to move in. Until March 1985, capital investment in industrial and commercial buildings attracted 100% tax allowances. Subsidies have also been directed at firms involved in specific projects through the Urban Development

Grant (UDG) and its Scottish equivalent, LEGUP (Boyle 1985). These grants have been used to support conversion, improvement or redevelopment schemes in which a substantial proportion of the cost is met by the firm itself.

Indirect subsidies have also been used. For example, local authorities have reclaimed land to make it suitable for redevelopment at no cost to the private sector. They have also granted licences to housebuilders to develop publicly owned sites for private sale. Thus site acquisition costs have been reduced or eliminated, creating greater profitability for development schemes that otherwise might not have attracted private investment. Many low-cost home-ownership schemes have been launched in the 1980s on the basis of these forms of hidden subsidy (Forrest *et al.* 1984).

The prime example of leverage planning in the 1980s is the London Docklands Development Corporation (LDDC), which we examine in detail in a case study. As its Chief Executive has explained, its primary objective is to generate private-sector investment (House of Commons 1984, para. 651). The establishment of the LDDC and its Merseyside counterpart was followed in 1987 by the designation of five further urban development corporations (UDCs).

The second batch of UDCs are Trafford Park (near Manchester), Cardiff Bay, Tyne and Wear, Teeside and a Black Country UDC based in Sandwell and Walsall. These UDCs have each been promised about £160 million of public investment over a five-year period. Their aim will be to stimulate a dramatic increase in the level of private-sector investment in their areas. None is expecting the scale of private-sector response seen in London's Docklands, but each is confident that development proposals can be attracted to their area. A third wave of UDCs was announced later in 1987 and in early 1988 covering Central Manchester, Leeds, Sheffield and Bristol.

The principal features of all these examples of the leverage planning style are first, subsidies to private-sector development, either directly through low-cost land sales or indirectly through infrastructure investment; and secondly, a flexible, even entrepreneurial, attitude to development proposals. The approach carries with it a strong implied criticism of the past record of local authorities in dealing with problems in these areas. Instead, emphasis is laid on the potential of the private sector to solve the problems, if only they had more confidence in the area. Considerable effort and money is therefore expended on boosting such confidence and improving the area's image.

This style of leverage planning is, as Young (1985) points out, highly interventionist. Although the political rhetoric behind this approach has emphasized the role of the private sector, in practice it depends on strong initiatives and a very active role by the public sector. Public officials are required to develop contacts with private-sector agencies and, in many cases, to put together a complete development package to be sold to private-sector investing institutions: 'This is not an arm's length activity. It is a "hands-on" interventionist approach' (ibid., p. 21).

Public-investment planning: directing urban change

There are many examples of this style of planning in the postwar period. Comprehensive Development Areas, first proposed by the 1945–51 Labour government as a mechanism for dealing with bomb-damaged areas, were envisaged as planning by public investment. The state purchased the land, compulsorily if necessary, and undertook most of the redevelopment. Large areas of Birmingham, Coventry and other cities were renewed in this way (Ravetz 1980). Public investment was therefore seen as a means of dealing with severe dereliction.

The most widely praised example of the ability of the public sector to plan urban change by investment is not, however, found in derelict areas but in the New Towns. Instigated by the New Towns Act of 1946 and developed from the much earlier ideas of the Garden City movement, this was the lynchpin of many policies: an instrument of regional planning; an example of sensitive local planning; a location for overspill council housing; and the counterpart to urban containment policy. The successful implementation of New Towns rested on a massive public-sector investment programme, providing urban infrastructure as well as services and urban facilities. The whole was coordinated by a public-sector plan, within which the private sector played a distinctly subordinate role. Direction remained the prerogative of the state, even in later years of the programme when more private investment was brought in.

More recently, public-investment planning has been specifically directed to the rescue of derelict areas. An important example of redevelopment by public-sector investment is the Glasgow Eastern Area Renewal project (GEAR), initiated in 1976 as the successor to the proposed Stonehouse New Town. GEAR forms the subject of our case study of public-investment planning. In this case local authorities, public-sector infrastructure bodies and the Scottish Development Agency (SDA) have together put substantial funds into rebuilding one of the most run-down areas in Scotland. The public sector has funded and coordinated the proposals for change.

Given the dominance of New Right ideologies at national government level, it is perhaps surprising to find that there are still localities in which planning by public investment is being strongly promoted. The initiative has come from some left-wing Labour local authorities, who have made an attempt to rebuild inner areas on non-market principles. The enterprise boards and Sheffield City Council are perhaps the main surviving examples (Boddy 1984, Wainwright 1987). Although there are significant differences between them, all of these bodies have attempted to redirect publicly controlled funds to support their local economies. In addition, they have used the considerable purchasing power at their disposal to maintain local employment.

Underlying these policies, and public investment planning in general, is the view that the British economy has been weakened by the investment plans and priorities of private-sector financial institutions – the banks, insurance companies, building societies and pension funds. Deindustrialization, decentralization and the resulting desolation of many urban areas is seen not as an accident of world economic trends but as a direct result of investment decisions by the agents of finance capital (Community Development Project 1977). It is therefore considered unrealistic to expect these same agents to provide the necessary investment to rebuild the inner areas. Only by the public sector taking over this role to direct socially useful development can the economic base of these areas be rebuilt.

This style of planning therefore exhibits great faith in the public sector, given comprehensive planning at all levels, good coordination between levels and adequate funding. Planners within this style 'network' in order to gain resources and implement strategies. The goals that can be achieved are similarly comprehensive, covering housing, employment, social welfare and regional balance, amongst others. The potential of such total planning for an area still attracts many on the Left. The question mark which hangs over this strategy concerns the ability of area-specific and limited investment resources to counteract the trends set in motion by huge flows of finance under the control of private institutions.

Private-management planning: handing over to the private sector

From the perspective of the New Right, the recovery of the most deprived and run-down areas of our towns and cities ought to be achieved not by massive state intervention, but by handing over the management of the whole renewal process to the private sector. This goes well beyond leverage planning, and draws in not only private-sector financial resources but also the managerial methods, skills and experience of the private sector. Its dynamism, creativity and energy, it is argued, can be harnessed to pull the run-down areas up by their boot straps, with the co-operation of local people and businesses (Heseltine 1983, 1986).

In the early 1980s some policy advisers have gone even further and proposed a new type of private-sector managed and funded city development agency to bring deprived inner areas up to national standards in housing and employment. They argue that it is only by the private sector taking such areas into its care that the processes of renewal and recovery can be made to work (Moor 1984, Henney 1985). A number of *ad hoc* initiatives have supported this vision of a new role for the private sector. The disposal of council estates to private developers for renovation and resale has encouraged the idea that private agencies are able to take over such areas. In a few cases the process has been extended to private-sector involvement in the management of renewal, through the mechanism of a non-profit-making private trust. Stockbridge Village, Knowsley, provides our case study and is a key example. The former GLC estate at Thamesmead has also been taken over by a trust.

A similar growth in private-sector involvement can be observed in the economic development field. The Community of St Helen's Trust, for example, was founded in 1978 as a result of the concern of Pilkingtons, the glass makers. Other initiatives and trusts, involving major companies such as Marks & Spencer, GEC and IBM, have followed. In 1982, with the government's encouragement, 'Business in the Community' was formed to promote such enterprise trusts and agencies.

Young (1985, 1986) identifies these ideas and initiatives as part of a broader strategy of privatization adopted by the Conservative governments since 1979. Under this strategy, private-sector agencies take on tasks that were previously seen as the exclusive responsibility of the public sector. Since these tasks are pursued within a broad framework of government policy, what is achieved amounts to the private management of public policy.

An important question is on what basis can the private sector be persuaded to undertake such a role? After all, the deprived inner areas are precisely those locations which the private sector has abandoned. Young argues that the Conservatives are in fact involved in 'a long-term attempt to persuade, cajole and tempt companies into believing they have a responsibility ... to the community at large' (1985, p. 26). For reasons of improved public relations, or out of genuine concern, some private companies may well be prepared to undertake some special projects of this nature. But it is the belief of some Conservatives that it should be possible to establish a broader corporate responsibility among major private-sector companies. The aim is to obtain a recognition on the part of business that it cannot simply opt out of concern for the social and economic problems of the country.

Another feature of this form of private-management planning is the extent to which public-sector subsidies, either hidden or more openly provided, are likely to be necessary for the success of projects. The way in which public-sector resources appear to underwrite the private-management planning style emerges as a major theme in our case study of the Stockbridge Village Trust. Such public-sector subsidy is also built into the Housing Action Trusts which were announced after the Conservative's election victory in 1987 and embodied in a White Paper, *Housing: the government's proposals*, of the same year (DoE 1987).

From typology to case studies

In this chapter we have identified six styles of planning and the essential characteristics of each style have been discussed. Although aspects of all six styles can be seen in British planning debates before the 1980s, we are arguing that together, in their current forms, they constitute a turning point in the history of planning, with important implications for both the practice of planning and urban policy generally. This reflects our belief that planning experienced a crisis in the 1970s. The debate which encompasses these six styles is part of the process by which that crisis is being resolved. The current fragmentation of planning into a number of distinctive styles has resulted from accelerated economic restructuring, which has heightened the contrast between local areas, and from a growing polarization of political ideologies, which has emphasized the contrasts between left- and right-wing attitudes to planning.

The purpose of the case studies, presented in the next six chapters, is to see how far each proposed planning style constitutes a coherent and distinctive approach in practice or if, instead, they differ more in their rhetoric than in their substantive effects. In doing so we will be moving from the arena of ideology and debate over planning solutions to that of policy practice, its formulation and implementation. Such practice is often chaotic and its effects unanticipated. This is partly because the policies themselves are not coherent, and each proposed style contains inherent contradictions. Partly it is because policies are being implemented in circumstances in which local agencies are forced to engage in opportunism, experimenting with many policies in the competitive search for resources, public and private. The result is not straightforward. It is not to be expected that any planning style is unproblematically implemented. This is particularly the case if one accepts that part of the function of proposing certain styles is ideological.

Case studies provide an ideal opportunity for exploring the styles as they operate in practice. Each case study allows in-depth analysis of a style and, in particular, of the interaction between agents that constitutes the practice of planning. They are the most appropriate method of studying processes, in this case the processes that characterize planning practice in the Thatcher years. In each of our case studies these processes are examined in relation to three key issues: the institutional arrangements of the style; the politics and mode of decision-making adopted; and the resulting conflicts and tensions. The distinctiveness and key characteristics of each of the planning styles becomes even more apparent as attention is directed towards these issues.

3

Regulative planning: the Cambridge area

Regulative planning has been identified by us as a style of planning appropriate to buoyant local economies. In such circumstances, it is argued, it should be possible for the public sector to harness the energies of the private sector, to divert and influence development in the knowledge that potential profits are high and marginal developments are few. The key to control lies in providing restricted development opportunities in certain locations and in exercising a veto of development in other areas. In this way concessions, in the form of a share of development profit, may be won for the local community. Individual private-sector developers accept such control in order to get some development rights, although there may also be more general benefits to the development industry in, for example, limiting competition between developers and providing an orderly pattern of development.

Regulative planning is the public sector making full use of the powers of development control contained in the 1947 Town and Country Planning Act. The local community benefits from protection of certain areas from any development, and the assessment of permitted development in terms of goals and criteria set by the community. The goals themselves are agreed through the participatory aspects of the planning process and the operation of representative democracy within the local council.

The Cambridge area

Examining regulative planning in practice requires a case study of a strong local economy since a high degree of private-sector interest is the prerequisite for such planning control. The Cambridge area provides an excellent example, for it is one of the local economic success stories of the 1980s (Fig. 3.1). The region of East Anglia extends eastwards to the North Sea and covers large

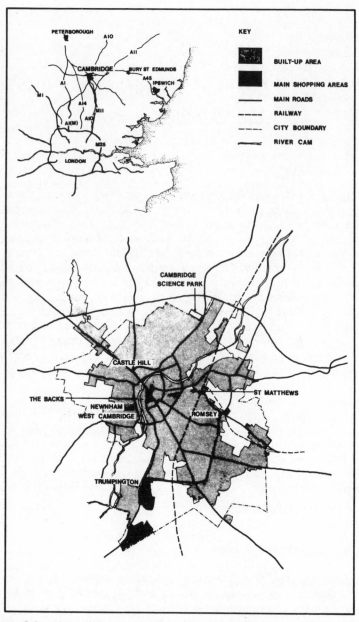

Figure 3.1 Map of the Cambridge area, showing places referred to by the case study

areas of agricultural land. Rural poverty is a problem over parts of the region, but its very rurality has allowed East Anglia to benefit from the shift in industrial location from urban to more rural areas (Fothergill & Gudgin 1982). More importantly, the region has the dynamic growth poles of Cambridge and Peterborough on its western edge. The percentage growth figures are therefore impressive, although it should be remembered that they are measured from a low base level.

In the mid-1970s, East Anglia was the only region to record an increase in manufacturing employment. By the period of June 1979 to March 1983 the recession had begun to bite but East Anglia, along with the South East, had the lowest regional levels of employment decline (5.7%) in the country. The percentage decline in manufacturing was lower than for the South East, and it was the only region to register an increase in service employment over the period. In the more recent period of March 1983 to September 1985, East Anglia had the highest percentage increase in total employment (7.9%) and in both manufacturing (2.9%) and service (12.2%) employment (Martin 1986). Such relative prosperity has drawn high levels of in-migration, with the result that population growth for the region has also been greater than for other regions: 13% during 1961–71 and 12% during 1971–81.

Within the region, the picture for Cambridge and the surrounding area is even rosier. The city is 55 miles north of London, 25 miles from Stansted and benefits from the M11 and two rail services into London. Unemployment rates in Cambridge are low, at 7.4% for the city and 5.1% for the Cambridge area as a whole. Cambridge's population grew by 11.6% during 1961-71 and then by 14.4% during 1971-81. The pressures for growth are likely to continue with the expansion of Stansted, the electrification of the route into Liverpool Street (which will reduce the journey time to under one hour) and the possible electrification of the other route into Kings Cross. Furthermore, the local economy is experiencing strong indigenous growth. The purchasing power resulting from these trends, together with over £100 million spent by at least three million tourists each year, makes Cambridge a major retail centre and pushes retail rents up to high levels.

The current dynamism of the Cambridge local economy undoubtedly rests on the so-called 'Cambridge Phenomenon', the growth of high-technology industry in the area. 'Silicon Fen' ranks alongside 'Silicon Glen' and the M4 corridor as one of the specialized localities where the new post-industrial revolution is occurring. The Cambridge Science Park opened in 1973 and since then there has been considerable growth in high-tech firms (Carter & Watts 1984,

Segal, Quince & Partners 1985). The pressures of this growth and the associated planning potential are the major issues facing local authorities in the area.

Regulative planning in Cambridgeshire, 1945–76

Local planning in and around Cambridge has a history of tight control of development and, unusually, has enjoyed fairly consistent ministerial support for strong regulative planning. The key to this history lies in the importance of the University (here used to include colleges, faculties and the University) in local politics, in local land ownership, and in local employment. Throughout the postwar period, local planning has been based on the premise that Cambridge should remain predominantly a university town. During the years 1931–48 it seemed as if industrial growth was going to eclipse the University. By 1948 there were three industrial operatives and two public-sector employees for every University employee (Senior 1956). The spectre of Oxford was raised; Cambridge, it was felt, should not allow industry to develop as Cowley had developed in Oxford. This principle was built into the first plans prepared under the 1947 Town and Country Planning Act.

In fact, the benefits of planning control had already been seen by the local council. As early as 1925, a town planning scheme had been prepared for East Cambridge. Planning activity within Cambridge City Council continued right through to World War II. Then, in the first of several ministerial interventions, the Minister of Town and Country Planning, Lewis Silkin, proposed that a planning consultant be called in. The result of the ministerial proposal was the 1950 Holford Wright Plan. This implied limiting future growth to a maximum population of 100 000 for the city, resisting industrial development and, in particular, discouraging any form of mass production. The 1954 County Development Plan, including Town Map 1 for Cambridge, was based on these principles, as was the 1957 Town Map 2 covering the ring of villages around Cambridge. Growth was to be dispersed beyond these villages, which would only take limited additional population. The Department of Trade and Industry supported this policy in its attitude towards issuing Industrial Development Certificates.

The first review of Town Map 1, approved in 1965, continued tight restraint and spelt out the policy in more detail. Only new industries employing five or fewer people were to be permitted, together with moderate expansion of existing firms. New industry linked to the university, i.e. 'science-based industry' and research and development activities, would be viewed more favourably. The success of these policies in the University's terms was evidenced by the fact that, in the period from World War II to 1966, the university population grew five times as fast as the city population.

Throughout the 1960s, in common with development plans elsewhere, these restrictive policies came under pressure from population and economic growth. The City Council was also concerned at the implications of such tight restraint for the growth of Cambridge and the employment prospects for local people. Their 1966 document *The Future Shape of Cambridge* therefore proposed increasing the limit on population growth to 120 000. The 1971 County Development Plan Review was also more relaxed in its attitude, particularly towards science-based industry. Furthermore, the Parry Lewis Report on the Cambridge subregion, published in 1974, called for a new settlement just outside the city to the south, based around a hypermarket. Growth was seen as inevitable, the result of general economic trends, but authorities involved in planning at all levels were concerned to control that growth, both in magnitude and location.

In the event Parry Lewis's proposal was rejected. Instead, three residential sites to allow for the extra 20 000 people were allocated in the 1976 Development Plan Review for the Cambridge sub-area. Restraint on industry continued and no new office development was allowed in the central area. Elsewhere, potential office developers had to demonstrate a link with the subregional function of Cambridge. This 1976 document was adopted by the County Council but not as a statutory plan; rather it formed part of the work on the first structure plan.

Structure planning

The *Cambridgeshire structure plan* was submitted to the Secretary of State for the Environment in 1978 and was approved, with modifications, in 1980. This document shows both the continuing concern of the county to control growth and its desire to achieve social goals through land-use planning. Three alternative strategies were put forward: conservation of the *status quo*; encouragement of economic growth; and an attempt to reduce inequalities within the county. The selected strategy combined social objectives with an element of conservation, a potentially conflicting combination (Healey 1983, p. 13). The four stated aims of the structure plan were:

31

(i)   the adequate provision of jobs, services and facilities with priority for the existing population and its natural increase rather than for the needs of the incoming population;

(ii)   the improvement of the quality of life in those parts of the county and for sections of the community which are relatively disadvantaged;

(iii) the protection of high quality agricultural land and a reduction in the rate of consumption of non-renewable resources of all kinds;

(iv) the conservation and improvement of the urban and rural environment. (*ibid.*, p. 20)

In practice, this involved following the 1974 report of the Regional Strategy Team, *Strategic choice for East Anglia*, in seeking to divert growth from the south and west of the county towards the Fenlands areas in the north and east. A degree of restraint on industrial and office development in and around the city remained, and local links had to be demonstrated before development was permitted. The intervening six years to structure plan review have brought surprisingly few changes in intention. The review was scheduled for an Examination in Public in late 1987 and approval in early 1988.

The aims are almost identical to the 1980 document. The strategy still seeks to divert growth within the county and argues that 'the potential for economic development and employment growth in the county should be fulfilled within a positive guiding framework'. Industrial development is still restricted to scientific and R&D activities and small firms. Speculative office development is discouraged. Growth pressures in the 1980s had led to previous targets for residential development being breached, and it was feared that considerable land would have to be allocated for new housing. However, changed population forecasts and negotiation within the County Council resulted in a downward revision of the figures. Instead of two new settlements outside the city to accommodate growth, only one is now proposed. Altogether, the structure plan proposes an extra 54 400 new dwellings in the county over the period 1986–2001. Some 15 200 dwellings will be required in the Cambridge sub-area.

Local plans, development briefs and planning gain

It has to be said that the rhetoric of control in structure planning is not unusual, though Cambridgeshire has had more support than most other counties for a policy of restraint. The test of regulative planning lies in the effective implementation of such policies. This section examines the implementation of policy through local plans and development control at the district level.

In the Cambridge area this falls to two local authorities: Cambridge City Council and South Cambridgeshire District Council (South Cambs. District Council). The reorganization of local government in 1974 left the boundaries of the city tightly drawn with the surrounding rural area, the developing suburb of Cherry Hinton and various villages all falling within the South Cambs. District. As will be explored later, this can give rise to political and organizational conflicts but, on the principle of effectively controlling development pressure, the two councils are largely in agreement.

Until recently, both the City Council and South Cambs. District Council have relied heavily on the statutory development plans, the 1965 and 1957 Town Maps, for detailed development control guidance (supplemented by the broader policies of the structure plan where relevant). District plans have so far been used on an *ad hoc* basis, although South Cambs. District Council is now preparing a district-wide plan. To date, the City Council has prepared three district plans: for St Matthews (1977), Newnham/West Cambridge (1984) and Romsey (1986) (Fig. 3.1). South Cambs. District Council has also prepared three: for the villages of Waterbeach/Landbeach (1983), Sawston/Pampisford/Babraham (1984) and Milton (1985).

The Romsey and St Matthews plans cover areas of later-19th-century terraced housing, some lower-density interwar housing (mainly in Romsey), local shopping and various industrial uses, often within the residential areas. Housing and environmental improvement are the main issues. Implementing improvement policies in the early 1970s had not been entirely successful. The City Council's approach had alienated many local residents. Following the recommendations of the 1969 Skeffington report, a more participative mode of dealing with improvement policies was therefore encouraged, and in 1975 the City Council published a manual for engaging in this sort of community involvement (Darlington 1975).

This clearly informed the district plan exercise in St Matthews. In addition to the more common programme of meetings with local residents, exhibitions and talks, the council undertook new initiatives. A working party of 16 residents, six ward councillors and planners led the development of policy; a 'planning shop' was opened in the area; and a newsletter was regularly circulated. On the basis of local involvement, two uncompleted Comprehensive Development Areas, which were causing blight, were abolished. A policy of enhancing the residential environment and of rehabilitation was adopted, and a General Improvement Area (GIA) approved in 1977.

The implementation of improvement policy, community development and the preparation of the district plan went hand in hand. It could be argued that the enhanced participation arrangements represent the first tentative steps towards the popular planning style. But the planning purpose was improved regulative planning, and the final document emphasizes the various powers available to the local authority to direct urban change: enforcement action, discontinuance powers, public health legislation, traffic management, and development control itself. The plan is now being updated with a view to making it a statutory document, and therefore more defendable at appeal.

The *Romsey district plan* covers an area adjacent to and similar to St Matthews, but the GIA here had been established prior to the district plan exercise of 1981-6. Whereas the St Matthews plan of the 1970s represents an attempt to move beyond the prevailing regulative planning style, the Romsey plan is very similar to the other four district plans mentioned, all prepared in the 1980s. Plan preparation was more conventional in its consultation with local residents and in document layout. The result was more formal and perhaps less 'user-friendly'.

The common link between the five plans of the 1980s is that each plan seeks to provide detailed guidelines for regulative planning in specific local circumstances. In particular they seek to exercise detailed control over new developments. The three South Cambs. District Council plans arose from the need to allocate specific sites following the approval of the structure plan. The Romsey plan had to deal with three key sites ripe for development, as well as resisting more general pressures for redevelopment and intensification in the area. The Newnham/West Cambridge plan was faced with intense development interest from the private sector, and a need to come to terms with the plans of the university, who own most of the land in the area.

The preferred strategy used in the plans is to include development briefs for particular sites. These set out precisely the requirements to be met before planning permission will be granted for the sites. The local authorities are here engaged in an exercise to influence development in order to meet community goals concerning the standard and nature of development. The district plan is used to advertise the council's development control powers and the concessions necessary for the grant of development rights. This is no open-ended promotion of a development site. Rather, the council sets the terms of negotiation on a planning application. This is, in effect, an invisible form of planning gain, since the planners are influencing the planning application before it is even submitted.

34

This form of influence is not normally acknowledged as planning gain. Instead attention has been focused on the more overt forms of planner influence (such as Section 52 agreements and concessions subsequent upon negotiation) and on the more substantial concesssions such as community centres and commuted car parking payments (Jowell 1977, Hawke 1981, Reade 1982). The routine exercise of power, affecting the preparation of proposals within the developer's office, should not be underestimated. These development briefs cover specifications for development density, dwelling mix, landscaping, access, parking, cycleways, phasing, children's play areas, soundproofing and standard of design – a comprehensive list! Even the planners do not necessarily accept this as planning gain but it is a clear case of controlling private development proposals in order to meet public goals.

More overt examples of planning gain can also be found. South Cambs. District Council uses Section 52 agreements (contracts between the applicant and the local authority conditional upon the grant of planning permission), mainly to deal with drainage problems on sites to the north of Cambridge due to the lack of Anglian Water Authority investment, though they have also been used to maintain open space. On certain applications the committee asks planners to negotiate concessions but, given the limited scale of development allowed in the district, it is recognized that the extent of overt planning gain will be commensurately modest.

The City Council relies heavily on negotiation, e.g. to upgrade the standard of development through the use of higher-quality materials. Negotiation, in conjunction with the council's power as a landlord on various industrial estates, is also used to relocate non-conforming industrial users from residential areas. Section 52 agreements may reinforce the use of standard development control powers, including discontinuance and enforcement powers, in such cases. In addition, local user conditions are routinely attached to office consents in the city. In many parts of the city these powers are given additional bite by the large Conservation Area (extended in 1980) and the widespread use of Tree Preservation Orders.

The use of development control powers to extract planning gain, in whatever form is, however, dependent upon the level of development pressure. Where the pressure is not urgent, regulative planning loses its force. For example, the Newnham/West Cambridge Plan accepts the long established prior claim of the University to develop the vacant areas to the west of the city. However, the University can take a very long-term view when it comes to development of its land, and thus precise details cannot be laid down. For example, the University has had plans to develop the Old Addenbrookes Hospital site in the town centre since 1962 (Cambridge University 1962); contracts were only exchanged in 1985 and development of only part is currently proposed. This makes it very difficult to include detailed site-specific planning policy in the local plan (it currently contains only one, for a research park), and any negotiation must wait on the University to submit a planning application.

Local planning and the Cambridge Phenomenon

Regulative planning involves an attempt to fulfil certain objectives through a fairly clumsy set of tools: structure and local plans, and the legal powers of development control. The objectives may be clear-cut, as in the case of very restrictive exclusionary planning, or they may involve a more complex set of social aims including employment maintenance, design standards, relocation of non-conforming uses, economic balance within a county, etc. It is in the pursuit of these more complex aims that the limitations of a regulative planning style based on negative control of largely private-sector initiatives become most apparent. This is currently very evident in the Cambridge area, because of the substantial development pressure which is emanating from the high-tech industrial sector. This section considers the application of regulative planning in the context of the Cambridge Phenomenon.

Given the outstanding reputation of the University of Cambridge for scientific research, there have long been local proposals to link academics with appropriate industry. This was evident in the industrial policies of the earliest development plans. However, the Cambridge Phenomenon, the growth of high-tech firms in the area, is usually tied to the opening of the Trinity College Cambridge Science Park (Fig. 3.1).

First proposed in 1969, the Cambridge Science Park (CSP) had its first tenants in 1973. Letting was fairly slow during the 1970s but since then the park has became the focal point of high-tech growth. It currently covers some 82 acres with a fourth phase proposed on another 26 acres, leaving a further 18 acres for later expansion. By late 1984 there were over 40 tenants, all occupying purpose-built accommodation. As has now become the norm with high-tech 'campus' developments, site coverage is low at 20%, allowing room for expansion, with a very high standard of landscaping, design and materials. Strict control over the development has come from two sources. The owner of the site and the developer, Trinity College, has used restrictive covenants in the individual leases. In addition,

South Cambs. District Council, the local planning authority for most of the site, has used its regulative planning powers.

The site was intended to be within the green belt and applications for housing and industry had been rejected during the 1960s. However, a review of policy in the light of the 1969 Mott report by a Senate subcommittee led to its identification as suitable for a Stanford-type science park. Having accepted the principle of such development, the district council then used a Section 52 agreement to guide that development. The agreement mainly controlled the use of the buildings and landscape maintenance. The control over use was particularly important since the then-current Use Classes Order would have allowed other forms of light industry and/or speculative offices to establish themselves without the need for planning permission. It was therefore necessary to use a Section 52 agreement to extend the limits of planning control (Brook 1983).

The recent review of the Use Classes Order, which proposes a new business class merging light industry and office uses, is intended to deal with this problem. The Royal Town Planning Institute argues that this will still not allow planners to control high-tech development in the way the market itself would wish, identifying a possible market-support role for regulative planning. Representatives of developers seem most concerned that the new Use Classes Order should reassure local planners and councillors and thus encourage the granting of planning permissions in situations of current restraint. In Cambridgeshire a local attempt is being made to solve the problem by providing a definition of high-tech development in the structure plan review.

The development of the CSP is generally regarded as a success in strict commercial terms and in the scope of its economic impact. It has become the market leader for high-tech development in the area. Recent research has estimated that in mid-1984 there were over 300 high-tech firms in the area with a joint turnover in excess of £890 million (Segal Quince & Partners 1985). The majority of firms are young and small, with a low failure rate compared to other new, small firms and with a greater tendency to generate spin-off companies. They are usually specialized, subcontracting precision engineering and similar tasks. The Cambridge Phenomenon is now very much a self-generating one, but changes are on the way with the increasing presence of international companies and the rapid growth of a few firms to a much larger size.

The growth of the Cambridge Phenomenon has been attributed to a number of factors. The promotional role of the university (e.g. in allowing intellectual property rights to accrue to the individual) and the availability of finance (initially from Barclays Bank) were early stimulants. More recently, firms have chosen the Cambridge location because of family links, the pleasant local environment and the prestige of the address, thus echoing research on high-tech industry elsewhere (Macgregor *et al.* 1986). The network of information and personnel between the firms themselves is considered important in generating growth and spin-offs, as are contacts with the university, but the precise significance of a location near a higher education establishment is debated (Haugh 1986).

Within this list of influences, credit should be given to the role of planning policies themselves. Following meetings between the county and city planning officers and representatives of local high-tech firms, the policies have actively supported high-tech development proposals. The strict criteria applied in individual developments have helped maintain the prestige quality of this sector of the market. Furthermore, the restrictive aspects of planning policy applied to other development, particularly industry, have perhaps unknowingly set the context within which high-tech industry can flourish. In the 1960s IBM was refused planning permission to locate its European R&D laboratories in Cambridge; analysts have suggested that if planning permission had been granted, the Cambridge Phenomenon might never have happened (Segal, Quince & Partners 1985, p. 63).

Links can also be drawn between these restrictive policies and the high-quality residential environment, the low level of unionization amongst the local workforce, the associated low wages and, in general, the absence of mass-production industry which has allowed small firms to reach the 'critical mass' for profit take-off very easily. However, the current results of past restrictive planning are now putting considerable pressure on those same planning policies.

The list of existing or proposed high-tech developments is a long one. In early 1985 the City Planner estimated that over 300 000 sq. ft of high-tech and R&D accommodation was in the pipeline (*Chartered Surveyor Weekly* 21 February 1985). For example, opposite the CSP, Pine Developments have built the first phase of the 20-acre Cambridge Business Park. The old Chivers factory site at Histon is being refurbished as 'Vision Park'. The Melbourne Science Park is being extended. Castle Park, on seven acres adjoining the county council offices, comprises high-tech units as well as new council offices. The former Cambridge City football ground is being developed with 170 000 sq. ft of R&D facilities. Aside from these larger schemes, there are many smaller developments including even, for example, conversion of rural pigsties to high-tech units.

With central government encouragement in the form of DoE Circular 16/84 (DoE 1984a), the county and city have been strongly in favour of these trends. South Cambs. District Council, following its early involvement in the CSP, has been less enthusiastic, seeking to preserve the local environment even from this prestige form of development, and to maintain more general restraint. For example, on a 22-acre site opposite the CSP, St John's College proposed another science park, the Innovation Centre. The site covered South Cambs. District and Cambridge City territory and, like the CSP site, had been proposed for inclusion in the Green Belt. The City Council was happy to support the proposal but South Cambs. District Council resisted it and won an appeal, thus restricting development to seven acres away from the A10 frontage. However, in another example of the successful use of regulative planning, South Cambs. achieved the landscaping of this undeveloped portion in exchange for allowing car parking on a small part. Another example is the Camtech proposal for a 55-acre science park in the Green Belt near the CSP, which was rejected by South Cambs. District Council. This resulted in a revised proposal, including retailing, a park-and-ride facility and a greatly reduced R&D element.

The growth in the local economy has had repercussions outside the high-tech sector, on the housing, office and retail markets. Segal, Quince and Partners' report comments on the local housing market as follows:

> The housing market, both in the city and the surrounding villages, is under pressure; and the shortages are probably most evident in precisely those categories likely to appeal to professional and middle-senior management. (1985, p. 90)

To deal with this, current strategic planning policies aim to allow limited housebuilding for these professional and managerial categories. In particular, they are attempting to steer at least part of the required development into a new village. In doing so they are in fact following market trends, which seek to provide a high-quality environment for up-market housing in separate settlements rather than by extending existing ones. In response to this structure plan policy, private developers have proposed a number of new settlements of their own.

Crow Green is a new village of 4000–5000 dwellings at Papworth Everard. It is being promoted by the Nationwide Building Society, on the basis of the new land ownership and development powers that building societies have acquired. Scotland Park at Hardwick is being put forward by Trinity and Churchill Colleges as a self-contained community of 2200 low-density dwellings with a high-tech/office park. Recently, Waterfenton has been proposed along the A10 to the north of the city. The developers, the Erostin Group, plan to build 3000 dwellings together with high-tech development, leisure facilities, a country park and an hotel. Consortium Developments have also announced their interest in developing a similar new settlement along the A10. These and other similar proposals are currently receiving a strong impetus from the land supply side as over-capitalized farmers consider the sale of land following changed European Community agricultural policies.

Restraint policies applied to office development are also being stretched as support services for the high-tech sector spring up: accountancy, merchant banking and public relations. In mid-1985 almost 250 000 sq.ft of offices were under construction (*Chartered Surveyor Weekly* 21 February 1985). The City Council routinely apply local user conditions to office developments. However, this has not prevented speculative office development nor the movement of large firms into Cambridge from outside, as intended. A firm can establish a small branch office to become a local user and then apply for a much larger office development. Local office users can obtain planning permission for new development, vacating their old premises which then, having an existing office use, can be speculatively redeveloped or refurbished.

Yet another sector of the market is also threatening the local regulative planning style with the buoyancy of demand, and that is the retail sector. Currently, retail planning is focusing on how to deal with pressures for out-of-town shopping. Given that the county estimates that out-of-town retailing will be required in Cambridgeshire by 1990 , there are voices within the City Council which would favour a hypermarket within the city boundary. It could then be controlled by the City Council to some extent and the impact on the city centre mitigated. For this reason there is a possibility that favourable consideration will be given to a proposal by the major retailers Marks & Spencer and Tesco, for a 250 000 sq. ft out-of-town shopping, hotel and leisure complex at Trumpington, a village south of Cambridge but just within the city boundary. Here the developers are offering planning gain: a park-and-ride scheme from the site to the city centre, which would alleviate some of the city's traffic problems; recreational facilities such as a multiscreen cinema; and payment for part of the Southern Relief Road.

The regulative planning style

The operation of regulative planning in Cambridge has provided an opportunity to examine this style in the context of a buoyant local economy. The Cambridge Phenomenon has both provided the development interest necessary for effective regulative planning and threatened to overwhelm local planning policy with the strength of that interest. It is now possible to consider regulative planning as an operating planning style in terms of the institutional arrangements, the politics and decision-making, and the resulting conflicts and tensions.

Institutional arrangements

Regulative planning presupposes a firm policy basis for development control, including consistent written policies and, preferably, a mapping of development allocations. This cannot be provided within one local authority on its own. Rather, it is provided through a network of local authorities. In small part this is due to the need of local authorities to co-operate, or at least to consult with each other over policies for adjoining geographical areas. But in the main the pattern of relations between local authorities in regulative planning derives from the joint effect of the Town and Country Planning Act 1968 and local government reorganization. The former created a dual system of structure and local plans originally intended for unitary authorities; the latter created a two-tier system of local government. As other commentators have noted, the result has been a succession of conflicts between county and district councils over planning powers. More recently, the attitude of central government (evidenced in the Local Government, Planning and Land Act 1980 and the 1986 DoE Green Paper on development plans) has encouraged districts to seek a more dominant position in local planning.

Even though there is an underlying consensus in the Cambridge area over the need for restraint policies and strong regulative planning, there have been cases of conflict over specific policy issues and development proposals. In the early 1960s, when the pressures for further growth were becoming evident, the County Council and a local farmer initiated a proposal to build a private-sector new settlement for some 5000 people at Bar Hill just outside Cambridge. The City Council opposed the proposal on the grounds that the village was too close to Cambridge, adding to the congestion in the city centre but, since it was outside the city boundary, contributing nothing to the city rate fund (Parry Lewis 1974). In the event the proposal was approved in 1964 following an appeal, since when a chequered history of housebuilding has meant that the development is only just being completed after almost 20 years of construction (Potter 1986).

The structure plan review has generated similar conflict over the location of further growth. It was originally proposed that two new settlements should be designated. Although there was agreement

41

over the concept of new settlements and over the use of the structure plan as a vehicle for their identification, there was disagreement over the appropriate location. South Cambs. District Council objected to the original two settlements, which were both in this area. Following the downward revision of dwelling numbers, one settlement in East Cambs. District is now proposed. Debate over its location continues.

As has occurred elsewhere, designating Green Belt boundaries has also been a source of conflict between the county and districts. In 1957 the minister approved a Green Belt sketch plan based on the principles of the Holford Wright plan. The review of the development plan redrew the boundaries. Town Map 1, covering the city and hence the inner boundary, was approved in 1965. Town Map 2, covering the outer boundary, was never approved, as it was overtaken, first by local government reorganization and then by a study of the Cambridge sub-area. It therefore forms a 'material consideration' in development control decisions but cannot constitute a statutory Green Belt. The structure plan incorporated the principle of a Green Belt but, given the nature of any Key Diagram, its precise location was left unclear. The Secretary of State's modifications had, in any case, limited the width of the Green Belt to 3–5 miles, rather than approving the extension southwards to the Hertfordshire boundary.

Therefore, in 1981, the County Council prepared a Green Belt subject plan. As elsewhere, this generated conflict over whether district councils or the County Council had the planning responsibility for such detailed mapping of the Green Belt (Elson 1986, Rydin 1986). The plan was published in May 1984, a public local inquiry was duly held in 1985, and the Inspector proposed some 50 amendments. The various district councils were not satisfied with these amendments, but it appears that most councillors in all authorities would have accepted the Inspector's report on the basis of everyone being somewhat disadvantaged. However, the relevant county committee meeting that considered the report decided to proceed on a site-by-site, amendment-by-amendment basis rather than discuss the report as a package. Given their local responsibilities, this encouraged local councillors to resist each and every amendment that reduced the Green Belt. It was decided not to accept the Inspector's report. The various district councils then reverted to their original objections to the plan itself. It thus became impossible for the county to adopt the subject plan prior to the structure plan review as they wished.

As in other counties, the subject plan is now left 'on the table'. Within the city the statutory Green Belt is that defined in the 1965 Town Map 1 and the Newnham/West Cambridge district plan. The statutory inner boundary is therefore more tightly drawn around the town than would have been the case if the subject plan had been accepted. This explains the City Council's willingness occasionally to give planning permission on certain Green Belt sites. In South Cambs. District the only statutorily approved Green Belt outer boundary is in the 1957 sketch plan, a document which unfortunately has been lost. Thus Green Belt policy has, to some extent, to evolve *de facto*, through development control decisions which make use of the non-statutory subject plan.

The involvement of a network of organizations can sometimes frustrate the preparation of a sound policy basis. Delay in preparation can create the possibility of policy being overtaken by events, e.g. where central government steps in on appeal and takes a decision over the local authorities' heads. The inherent conflict can also lead to serious policy confusion. For example, one debate over residential allocations in the structure plan review has concerned a 900 dwelling site known as Clay Farm, to the south of Cambridge. South Cambs. District Council favour its release, as does the City. The County Council steering group dealing with the structure plan review made explicit mention of Clay Farm as a proposed housing allocation. The Planning Subcommittee, however, while debating the Green Belt subject plan, for a time proposed putting Clay Farm back into the Green Belt!

Some argue that the difficulties of preparing a clear, up-to-date policy basis for everyday planning decision-making are compounded by the absence of clear regional guidance emanating from central government (Nuffield Commission of Inquiry 1986). Currently central government imposes its view on regional growth in an *ad hoc* manner through structure plan modifications and even appeal decisions. Local authorities therefore have to anticipate the likely reaction of central government to their policy, particularly a restrictive policy. Their discussions reflect not only local preferences but an attempt to forestall possibly higher levels of growth imposed by central government. In order to maintain local control over growth, concessions are made to a perceived central government viewpoint at plan preparation stage. Such a situation may well accentuate the scope for disagreement at the local level, as compared with an attempt to implement a settled regional policy. Against this can be set the view that adding a regional plan to the current hierarchy of plans would extend the bureaucracy of plan making and hence exacerbate the scope for inter-organizational conflict, delays in plan preparation and confusion of current planning policy in an area.

An attempt to cut through these organizational tangles was made by the Nuffield Commission of Inquiry into town and country planning. This proposed the issue of National Planning Guidelines and an annual White Paper on Land and the Environment by the DoE. Regional versions of the planning guidelines would also be issued. The county and district councils would prepare their policy documents within this framework. District and local plans would follow current planning practice, but the county strategy would be less extensive than current structure plans while the county development plan would be more detailed (Nuffield Commission of Inquiry 1986). This well developed set of proposals contrasts with the DoE Green Paper of the same year on *The future of development plans*, which suggested the abolition of structure plans and their replacement by district-wide local plans. These government proposals have been temporarily shelved owing to the weight of the legislative programme for the current Parliamentary session.

Politics and decision-making

The formulation of planning policy does not only have to cope with inter-organizational conflict. There are tensions within the local authorities between the various key actors, namely the local politicians and professional planners. These are of two kinds: party political conflict between councillors, and conflict between planners and councillors. The political tensions are highlighted in the Cambridge area by the composition of the various councils.

In May 1981 the County Council became a hung council. Following the last county council elections in 1985, there were 29 Conservative, 26 Alliance, 21 Labour members and one independent member. This is reflected in the Planning Subcommittee as in other committees of the council. The structure plan steering group has equal representation of the main parties among its nine members. The implications of the precise balance of a hung council were appreciated in that the structure plan review was delayed until after the May 1985 elections.

The main party political conflict in planning comes over the scale of growth to be accommodated, with a part of the Conservative grouping vigorously resisting even the modest planned growth proposed in the review. This led to an attempt by the Conservatives to prevent the adoption of the draft review document, first by a series of wrecking amendments and then by a partial walkout at a council meeting. In effect the other two parties are mediating a basic conflict between pro-growth central government Conservatives and anti-growth local Conservatives.

Planning policy itself is developed by the small steering group with planning officer support and advice. The relationship between councillors and planners in this task is quite a complex one. On the one hand, planners have the authority of professional expertise.

They act as a channel for central government advice, in this case stressing the importance to the national economy of encouraging the Cambridge Phenomenon. In smaller and closed meetings, such as the structure plan steering group, they can speak more freely than in open council and, by initiating particular sessions with invited delegates, they can lead policy debate.

On the other hand, during internal reorganization the structure planning unit has been reduced from over 20 staff working on the first plan to only four working on the review (compared with the nine councillors who were meeting fortnightly during the first year of the review). Furthermore, there is a high level of planning expertise on the steering group, since many of its members were involved in the preparation of the first plan. There is, therefore, also scope for politicians to lead professionals (Healey 1983, pp. 53–5).

Turning towards internal decision-making in the two districts, consideration of development control as well as more strategic policy matters is involved. Until local government reorganization, the City Council was controlled by the Conservatives and the University held eight seats. The election in 1973, on the new boundaries, gave Labour power for three years, but 1976–9 saw the Conservatives again forming the largest group. Since 1980 Labour have been able to take their place. In May 1986 they (just) gained overall control of the council, only to lose it in May 1987. The City Council operates with an Environment Committee and a Development Control Subcommittee of only seven councillors. As is commonly cited elsewhere, planning is not regarded as the most party political committee (Nuffield Commission of Inquiry 1986). Nevertheless, conflicts do sometimes divide along party lines, and there is perhaps an increasing tendency for party politics to intervene in decision-making.

The City Council Conservatives describe themselves as more pragmatic, with a greater understanding of developers' priorities and of the need for development. Yet they strongly support the Green Belt with its implications for reduced development levels. They are also less enthusiastic about the use of enforcement powers, particularly on small businesses. The Alliance and Labour councillors are fairly similar in planning attitudes, which is not surprising given the fact that many Alliance councillors used to be Labour supporters. This involves a less sympathetic attitude towards developers where development in principle has been accepted and, paradoxically, a more permissive attitude to growth *per se* given a concern with local employment prospects.

These views are given expression in discussion of general policy for the city area, in examination of Green Belt boundaries and,

45

above all, in development control. Here the party political conflicts are supplemented by other tensions. First, the party political stance of a councillor may conflict with the local responsibility of the councillor as ward representative, a particular problem for councillors on the planning committees. Secondly, the city councillors have developed a very active role in development control, and disagreements between planners and councillors are not infrequent.

Some of these disagreements concern the basis of development control decisions. Planners emphasize the limited land-use planning considerations that legally must underlie each decision, but councillors may wish to take broader issues into account (Loughlin 1980). Sometimes the disagreement is more fundamental, and councillors may overturn the planners' recommendations, usually in order to refuse the application. Councillors generally resent being pressurized by planning officers, as they see it, into accepting the implications of negotiation by officers on applications, perhaps in the form of a Section 52 agreement. On occasion, councillors may press for the preparation of a detailed policy statement when the planners already feel under pressure from a lack of resources. In short, the city planning councillors energetically seek to establish their own planning decisions and priorities, sometimes in the face of contrary professional advice.

In South Cambs. District Council, by contrast, party politics are played down to the point where only a minority of councillors are elected on a party platform at all. Instead, the majority are independent councillors. This means that when planning matters, policy and applications, are discussed by the 20 members of the Planning Committee, local issues very much predominate. The local councillor is always invited to speak first on any matter relating to his/her ward and, by and large, favourable consideration is given to these local views as well as any from the parish council. The members are generally very protective of the environment, keen to maintain the agricultural basis of the district and generally conservative in planning terms.

Although there is often extensive discussion of aesthetic aspects of an application, perhaps following on from the recommendation of the local Architects' Panel, there appears to be less attempt to broaden the basis of planning decision-making. Disagreements with planners seem rarer than in the City Council, with councillors more readily deferring to professional opinion. This may be due to the absence of any strong party line to oppose the professional viewpoint. Yet in cases where there is strong local opposition, it can still be difficult to implement a planner-led strategy, given the committee's commitment to local views.

These differences, along with less frequent use of enforcement action, follow from the more restrictive policy adopted by South Cambs. District Council. By and large, development is discouraged. Giving any possible precedent for future development permissions is strongly resisted. Even specific land allocations in district plans are viewed warily by some, on the basis that they get taken up so rapidly. An up-to-date structure plan is seen as a firm basis on which to exercise restrictive planning. In this view councillors and planners seem agreed, so that planners can routinely lead decision-making. This situation may change as the agricultural interests within the council shift from a protectionist strategy to an attempt to release land from agricultural use in a changed Common Agricultural Policy context.

In the city, however, the councillors wish to use planning as a tool for directing and influencing development in pursuit of broader social goals. In doing so they often try to expand the remit of statutory planning and hence come into conflict with planners. Professional planning ideology is not always aligned with local attitudes to development, and decision-making occurs through more active discussion between planners and councillors.

In each case though, and including the county, it is an élite of planners with councillors that undertakes the role of decision-making. Together they seek to execute the various procedures of strategic plan-making and detailed development control. Decision-making involves councillors making judgments on party political grounds and out of concern for their particular ward. But, as we have emphasized, it also involves considerable discretion and the exercise of judgment, from skilled and knowledgeable councillors as well as planners. Both councillors and planners use the procedures creatively and stretch them to their legal limit in an attempt to achieve social goals. We therefore characterize this mode of decision-making as technical–political.

Conflicts and tensions: coping with the Cambridge phenomenon

The Cambridge Phenomenon would seem to be an unmitigated success story of the marriage of strong market demand with regulative planning powers. But some of the local implications of the high-tech boom are causing a rethink, particularly within the City Council. The first of these concerns the impact on the local labour market, or rather the lack of impact. About 17% of employment in the sub-area, some 14 000 people, is accounted for by high-tech companies. But the majority of this is highly skilled employment, often filled by in-migrating workers. Even within the high-tech sector there is some replacement of skilled staff by lower-cost research students.

The scope for reducing local unemployment amongst semi-skilled and unskilled workers lies not in high-tech industry itself but in any multiplier effects. Currently, only 20% of the local workforce are employed in manufacturing. Closure of the Pye television works, following their takeover by Philips, has further reduced local manufacturing employment opportunities. There is also a tendency for production spin-offs from high-tech companies to be located elsewhere, often outside the UK.

The City Council is extremely concerned about this and about the low level of wages that have persisted in the city despite the boom. It has a vigorous policy of industrial development and has promoted a number of industrial estates such as the 13-acre Clifton Road Estate, developed in partnership with Dencora Securities, and nursery units at various locations in the City. In the three-year period to mid-1985, the City was able to arrange accommodation for 188 industrial firms. But the role of these industrial estates is being affected as high-tech firms locate there. This has occurred on the Clifton Road Estate with the development of the 34 000 sq.ft Camtech Centre, operating with no local user conditions and only one-third site coverage.

Some bodies, such as the local Communist Party (1979) and the Cambridge and District Trades Council (1976), would favour broader encouragement of industry. But within the constraints of current policy, finding a site for, say, a large incoming industrial user would be difficult. The current policy is the only politically feasible one, but it is acknowledged that its impact is likely to be limited.

Related to the relative shift in employment structure is a massive appreciation in local housing values. The effects of the Cambridge Phenomenon are here augmented by the growth in commuting to London from the city. This clearly reduces the ability of local workers outside the high-tech sector to purchase housing and it is even creating problems for key high-tech personnel seeking to move to the area, as Segal, Quince and Partners noted. Restrictive planning policies help to maintain house prices but their relaxation is unlikely to contribute to a fall given the reluctance of housebuilders to cut selling prices; building rates are more likely to adjust downwards in such circumstances. However, as long as housebuilding is profitable at these price levels, the restraint policies will come under attack.

This poses a dilemma for planners. Do they risk stifling the Cambridge Phenomenon if they do not allow more housebuilding, particularly for professional and middle–senior management? Or will more development itself affect the phenomenon by reducing the

48

environmental quality of the area? The current approach adopted by the local councils is to control the scale of new development and to guide part of it to a new village, in line with market preference. But past experience shows that planning targets for population growth can be breached by a buoyant development sector and the prospect of central government stepping in to ease the restraint policy remains.

Local policies for retail development may also be threatened by pressure from the private sector. At the time of writing, three appeals for retail schemes north of Cambridge were under consideration. If these appeals are allowed then development proposals will lead structure planning, not vice versa. Similarly, the debate over the Marks & Spencer/Tesco development at Trumpington is an example of development pressure producing specific proposals, and development control decisions on those proposals leading strategic planning. Here the City Council may give approval to a development which is within the Green Belt in order to control the details through regulative planning powers. If it does so the structure plan review will have to accommodate such a scheme and adjust accordingly.

In general, regulative planning is much more successful in dealing with individual schemes than in implementing a strategic policy given the unanticipated impacts of development, the insistent nature of strong market demand and the limited powers of development control. In the end the exercise of regulative planning depends on private sector demand and thus the style essentially requires a mixture of simultaneously encouraging and discouraging development proposals from the private sector. It must be recognized that regulative planning has been particularly well supported in Cambridge because of the concerns of the University, a major local landowner and political force. Until 1973 the University automatically had seats on the City and County Councils. Even now it is an important consultee for local planners. Throughout, it has also been able to rely on a degree of central government support for its viewpoint. For much of the postwar period this has meant a restrictive planning policy, although there have been occasional battles over developing particular sites.

However, the pressures for development are now in substantial part coming from the University itself. It has been variously described by interviewees as a 'vigorous entrepreneur' and a 'pirate'. Where such a major landowner and local political influence seeks development then regulative planning faces great difficulties. For while regulative planning does provide some potential for using market pressures to meet social goals, it can also be used as a mechanism for maintaining property values and selectively releasing the development potential of sites. At the moment local regulative planning in Cambridge is maintained by the general agreement on the need for controlling the excesses of strong development pressure. But the underlying tension within regulative planning remains. This is derived from the attempts by various participants to achieve different goals. The difference between the University's and the city and district councils' views of regulative planning is becoming more and more apparent as the high-tech boom continues.

4

Trend Planning:
Colchester, Essex

In our typology we have identified trend planning as lying at the opposite end of a spectrum of planning practice running from regulative planning. It is not, however, merely a weakened form of regulative planning, the embodiment of the 'death of planning': it is not non-planning. Rather it is a distinctive style, which uses the tools of the land-use planning system to pursue particular goals. Trend planning seeks to facilitate private-sector development rather than control it. The resulting economic activity and change in the built form is regarded as the evidence of successful planning. This style considers that the public interest is best served by the development itself rather than by any planning gain that may be secured from the developers. The preferred pattern of land uses in an area is that identified by market actors rather than professional planners, and the latter are urged to be responsive to market pressures.

The justification for such planning is couched in terms of encouraging entrepreneurial activity and freeing the wealth-makers in society from unnecessary red tape. It is suggested by its proponents that this will lift the shackles of excessive planning control from developers who will then be able to lead that locality to greater economic prosperity. This presupposes, of course, a certain level of existing prosperity and associated development interest.

The flagship of trend planning is the proposal for Simplified Planning Zones outlined in Chapter 2. But examples can be found of localities in which the existing statutory planning system is currently being operated in line with the aims of trend planning. Examining such an area throws light on several issues. It indicates the transformed effect of the planning system where the local public interest is redefined in terms of market outcomes. It emphasizes the difficulties of achieving even a limited set of additional or alternative goals through a reliance on market indicators. And it shows the continuing local attachment to certain limited aspects of regulative planning, which in turn raises the question of whether the time is yet ripe for re-orientating land-use planning any further towards the private sector perspective.

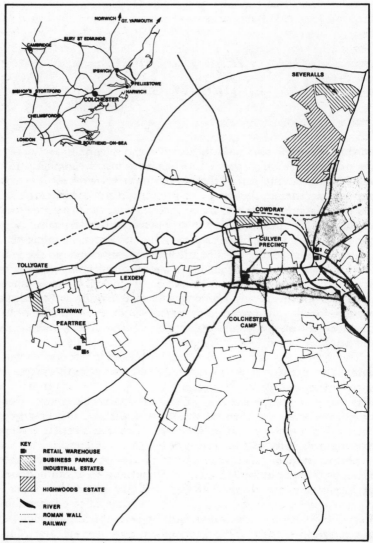

Figure 4.1 Map of the Colchester area showing places referred to in the case study

The Colchester area

The area chosen for this case study is Colchester in north-east Essex (Fig. 4.1). The main urban area comprises a market town of nearly

52

80 000 people (1981 Census) with a history dating back to Roman times and beyond. Many historic remains are evident, not least the Roman wall, the Norman castle and a number of 16th-, 17th- and 18th-century buildings. The borough of Colchester, incorporating several surrounding villages and much agricultural land, contains a population of 143 000. It lies at the outer edge of the relatively prosperous South East region. The town is 50 miles and 50 minutes by train from the City of London and has good contacts with the ports of Harwich and Felixstowe, and with the expanding airport of Stansted as well as the various towns and cities of Essex and East Anglia. The M25 and M11 have improved its accessibility, as will the upgrading of the A12 route to London. These advantages, combined with patterns of urban decentralization in the South East over the past two decades, have made Colchester an area of growth. During the 1961–71 period Colchester's population increased by 26% and during 1971–81 it increased by 13%. These rates are above the average for Essex and are largely due to in-migration to the borough.

The relative prosperity of Colchester is evidenced by the fact that the local unemployment rate is below the national average at 8.7% (July 1987). In addition to local manufacturing and agricultural industries, the economic base of the town is being diversified by relocating office users and other service-sector growth. The army camps to the south of the city also remain an important influence.

Facilitating development and influencing design

The extent to which planning in and around Colchester is based on encouraging private development is now recognized at borough, county and even regional level. In the past attempts have been made at a regulative style in structure and regional planning, but economic and political changes have led to a current convergence between planning policy at the different levels. This can be illustrated by looking at regional, structure and local planning in turn.

After several years in abeyance, regional guidance has recently been re-established in the South East following an initiative by the local authorities in the region (under the auspices of SERPLAN). The resulting document can only be described, at its strongest, as seeking to guide rather than control growth. SERPLAN (1985) argues that land for 600 000 additional dwellings will be available over the period 1981–91; this fits almost exactly with DoE 1981-based demographic forecasts for the decade. SERPLAN further proposes development of 460 000 additional dwellings in the region during 1991–2001; DoE demographic forecasts suggest 373 000 extra households.

The document rejects the traditional aim of regional guidance in the South East to divert growth to more depressed regions and instead seeks to foster economic growth within the South East. In his response to the SERPLAN document, Nicholas Ridley, the Secretary of State for the Environment, commended the 'realistic and pragmatic approach', stating that:

> At the outset it is necessary to recognise the limitations of the land use planning process. It is the private sector not the planning system that generates economic growth. But soundly based land use plans can help to facilitate development and investment... (SERPLAN 1986)

Structure planning in Essex also recognizes the leading role of the private sector. Indeed, for many years strategic planning in the county seemed to run after events. The first postwar development plan was quickly overtaken by the actual rate of development in the county. As a result the review of the development plan, prepared in the 1960s, had to come to terms with the large amount of land that was already developed but not previously allocated. This review was not even approved by central government until the mid 1970s. By that time the first structure plan was in preparation. Submitted in 1979, this was approved in 1982, but by then the anticipated decision on Stansted and the development of the M11 and M25 had rendered it out of date. Work began immediately on a review, published in draft form in 1986.

A significant theme in this latest document is the promotion of development. The structure plan no longer tries to guide employment within the county, removing any 'restrictive' element in its employment policies. It suggests allocating additional land for offices and industry but emphasizes that 'the totals, however, are not inflexible' (Essex County Council 1986, para. 3.2.8). With regard to shopping it notes that past structure plan policies had already been flexibly implemented, and specific floorspace targets for shopping in various parts of the county are deleted. Throughout the document there is an emphasis on flexibility, on encouraging development, and a recognition that the outcome of the plan depends heavily on market conditions and private developers.

This product of the Conservative-dominated County Council is very much in line with Colchester Borough Council's approach, which is described in property and development circles as 'progressive' (*Chartered Surveyor Weekly* 29 May 1987). The Borough Council has prepared two local plans, one for the city area and one for the rest of the borough. In these, six Areas of Development Opportunity are designated. These have been generally advertised as sites where a variety of development projects would be considered appropriate. While the local plan talks of the need to plan these sites comprehensively and of the potential for providing community facilities as planning gain, it recognizes that:

> ...the role of the private sector is likely to be crucial to the redevelopment and improvement of these sites. The rate of development will be dependent upon the economy and private investment. (Colchester Borough Council 1984, para. 12.19)

This view is backed up by the economic development activities of the Council. The borough's 'industrial and commercial advisor' seeks to market Colchester as a location for private investment. This is done generally, using promotional literature for the town, and for specific sites. The industrial and commercial advisor has made it clear that the prime aim is to attract private investment and encourage growth (*Chartered Surveyor Weekly* loc. cit.). To aid the council in this aim, a report has been commissioned on the economic potential of the outer South East subregion. The council has also appointed a Tourism Officer to encourage appropriate ventures. More specifically, the Council is engaged in marketing units on the 168-acre Severalls Park Industrial Estate, a development on land owned by the Council since the 1930s.

In the area of housing, the Council has long sought to run down its public housebuilding programme and rely on private developers, with only limited housing association involvement. In 1977 the council announced plans to end all Council housebuilding by 1981. In the face of a worsening local housing crisis this has not been carried through, and about 130 dwellings has been the average annual housebuilding rate. However, the new-build section of the direct labour organization has been disbanded. Recently, 120 surplus army dwellings at Lethe Grove were sold off to the private sector for refurbishment rather than being taken on by the council.

Given the limited extent of either strategic planning or direct public-sector involvement in development, effective planning policy is very much focused at the level of development control. Furthermore, the generally positive attitude towards development means that it is the details of the proposed design that receive most attention. The concerns of the Senior Development Control Officer and the discussion at Planning Committee over applications concentrate heavily on aesthetic and architectural matters. Concern with townscape, with the environmental quality of buildings and the need to replace 'scruffy buildings with attractive buildings' is seen as being 'what it is all about' (quote at Planning Committee).

The practice is to provide adequate land for development, to avoid head-on conflict with developers over the principle of development, and therefore to concentrate on improving the quality of that development. This may involve, in certain cases, overcoming local opposition to the principle of change. The success of the policy is gauged by the generally small number of appeals and the extreme rarity of losing an appeal on architectural grounds. Overall, though, the achievement of broad social goals is not sought and even the role of land-use planning in spatial coordination may be overlooked. The role of the public sector is a limited one, involving intervention in what many people would regard as peripheral rather than central issues.

The operation of this distinctive local style of planning can be seen in relation to the major development pressures in Colchester, for city centre redevelopment, new retail schemes and housebuilding.

City centre redevelopment

Redevelopment of the centre of Colchester has focused in recent years on plans for the Culver Street area (Fig. 4.1). Culver Street lies adjacent to the main shopping areas of the High Street and the Lion Walk pedestrian precinct. Like Lion Walk, Culver Precinct was allocated for development in the 1969 *Town centre plan*. The site had been largely vacant since before the war, when the council bought a factory on the site in order to relocate its occupants. After the war it was used as a car park but, once alternative parking facilities had been built, it was proposed that a new town square with surrounding shopping be developed on the site. It was felt that such a square would be a community facility as well as an attraction to shoppers and visitors from outside the area. With substantial council ownership this seemed a feasible proposal.

The first plans were drawn up by 1970 but the economic uncertainty of the early 1970s delayed progress, and changes in the retail market meant that design alterations were necessary. By the end of the 1970s, though, the names of C&A and Debenhams were being cited as the anchor stores. Developers interested in the project were asked to submit outline proposals by early 1979 and six duly did so. Detailed designs were requested by the end of the year. Two issues were raised in consideration of these designs: the size of the town square and the fate of a public library on the site. The latter was a 1930s building by Marshall Sissons, winner of a 1936 architectural competition. Though it was never completed according to the design, it was a popular Colchester landmark. However, from a developer's viewpoint there were difficulties in incorporating the building in the new precinct owing to the high ceiling heights.

At the time of shortlisting, the Borough Council wished to demolish the library and have a town square of 18 000 sq. ft. Many had hoped for a much larger square with the public library (probably in an alternative public use) as the central feature. Essex County Council, the library's owners, backed its retention but the Vice-Chair of the County Council working party on the library recognized that:

> ...in the current financial climate we must consider the ratepayers' interests first and foremost. (*Essex County Standard* 30 March 1979)

In the event, the two shortlisted developers, Carroll and Guardian Royal Exchange (GRE), both intended to keep the library and included a town square in their proposals. After another ten months, Carroll was selected as the developer and detailed negotiations began.

At this stage the vulnerability of planning based on private-sector actors becomes apparent. In Spring 1981 it was announced that Debenhams were pulling out of the scheme, reputedly unhappy with their proposed location away from the existing main thoroughfares. Two other stores who were approached declined to become involved, and the lack of an anchor store seemed to threaten the whole scheme. The prospect of the redevelopment coming to a halt silenced many previous critics of the scheme and instead led to calls for speedy decision-making by the council. It also no doubt influenced the Council's refusal of planning permission for a major shopping centre at Colne River Stadium just outside the town centre.

By mid-1982 the plans had been changed to accommodate Debenhams, although the store refused to commit themselves absolutely to the scheme. The design changes were not made public until pressure was brought to bear by opposition councillors. When they were publicized it became apparent that in order to improve pedestrian flow and profitability the town square no longer had an open aspect and was no longer focused on a public building, and that the library might be demolished. The number of shops was also increased from 13 to over 40. Public reaction was adverse.

There was also debate within the Council over the likely financial return the Council was to receive. As major landowner (and eventual site-assembler) the Council was a party to the development. At this time (early 1980s) it was reported that it had been offered the choice of a £6.25 million lump sum or an annual ground rent of £250 000. Some councillors had argued for more money, but the developers wanted the final design details of the development agreed before discussing financial shares of the profit. The Borough Planner had made it very clear that prolonged negotiation was not desirable, commenting that 'the longer the development takes, the less you are going to see for your money in the end' (*Colchester Evening Gazette* 14 November 1979).

Over both financial and design negotiations, opposition councillors argued that satisfactory compromises had not been achieved, as the Conservatives claimed, but rather that the council had capitulated to the developer's wishes. In August 1982 the chairman of the Property and Development Committee had to fight a no-confidence motion over his handling of the project. He survived to warn in early 1983 that the scheme was again on the verge of collapse if a design was not agreed:

> If we don't accept this scheme on Wednesday, we shall have a hole in the ground in Culver Precinct for many years to come. I just feel that we mustn't be too greedy. (*Essex County Standard* 18 March 1983)

In mid-1983 a planning application was finally submitted; it was approved in the autumn. Until the last minute there was a threat of Debenhams pulling out and jeopardizing the whole project. The final deal between all parties (the council, the developer and Scottish Amicable Life Assurance, the long-term financier) was not concluded until late 1985. At the time of writing (late 1987) the precinct is still under construction but has been partly opened.

At the end of the day, the town square is 22 000 sq. ft and the total shopping floorspace amounts to almost 248 000 sq. ft gross, comprising a department store, 33 unit shops and 8900 sq. ft of other small shops. The square might therefore be considered appropriate for a market town. However, it is a private square and although open to the public, is not to be used for any public demonstration or rally. The one completed wing of the Georgian-style library is being demolished, restoring the building's symmetry; the flat roof is being replaced with a pitched one and a retail store 'wrapped around' the façade. Last-minute design alterations have led to an atrium being incorporated into the Debenhams store, at the suggestion of the Senior Development Control Officer. The financial deal has been revised so that the council, reportedly, receives a lump sum of £0.25 million and the return, mainly in the form of a ground rent, of about £240 000 with 12.5% of the rental growth. The Conservative majority preferred this as they felt that a large lump sum would only create political pressure for 'inappropriate expenditure'. It will also help maintain the low rate levels that the Borough Council has achieved in the past.

For a time the Council also sought a return from its involvement in the development partnership in the form of new council offices. In the 1960s there had been plans to extend the 1902 Italianate Town Hall and centralize council operations, but this had foundered on ground condition problems. Local government reorganization had led to even more decentralized offices. Therefore in 1983 the possibility of council offices forming part of the Culver Precinct development was mooted. This was not a form of planning gain but a way of utilizing council land.

To cope with the Council's requirements, quite substantial offices were necessary. While such space could be provided fairly cheaply, since the council was the landowner, it reduced the direct monetary return the Council could expect from the development. The developer was not unhappy with the proposal since it involved no financial drain on profits and would provide a degree of assured pedestrian flow through the precinct.

Then in 1985 a local property company approached the council with an apparently low-cost scheme for council offices on Angel Yard, some Council-owned land adjacent to the Town Hall. This was enthusiastically received by the Town Clerk and ruling party, so much so that it was intended not to take the scheme to open tender. Opposition councillors were horrified, particularly when they discovered that the developer in question had made contributions to Conservative Party funds. This forced the Council to invite tenders, but by this time the idea of council offices at Culver Precinct had been dropped. This landlocked site remains without firm development proposals. Angel Yard, after a change of design to avoid demolition of some listed buildings on the road frontage and a dispute over ancient lights, should be completed in 1989.

Out-of-town retailing

Retail warehousing, superstores and other out-of-town shopping centres have proliferated in Colchester as elsewhere (Hillier Parker May & Rowden 1986) (Fig. 4.1). The *Central area local plan*, based on a 1980 land-use survey, identified four retail warehouses in the central area and another 12 in the suburbs. These outlets accounted for two-thirds of total retailing floorspace for bulk items such as furniture and carpets. Since 1980 more such stores have been developed. These changes in the retail market have led to substantial shifts in shopping facilities around Colchester.

An example is provided by the past and present location in Colchester of Sainsbury's, the supermarket chain. Sainsbury's opened their town centre store, with approximately 14 000 sq. ft of sales space, in 1969. Seven years later they announced plans for a supermarket in Stanway but were refused planning permission and at that time lost their appeal, with the Secretary of State overruling the Inspector's recommendation. In the same year that the appeal was lost, a local development company obtained planning permission from the Council for a 17 000 sq. ft supermarket in Lexden. The following year it was announced that Sainsbury's would take the Lexden site, though some believe their involvement had been more long term.

Severe problems occurred with this development as there was inadequate access and insufficient parking. More local opposition to the store arose with the possibility of Sainsbury's buying adjacent houses to demolish them for parking. Sainsbury's clearly needed alternative larger premises. In 1982 they proposed a superstore in the huge Colne River Stadium development. As we have seen this did not receive planning permission from the council, which was eager to secure the success of Culver Precinct.

In the next year the prospect of a Stanway superstore surfaced again, and later in 1983 the council altered its plans for the Tollgate Industrial Estate (since renamed the Tollgate Business Park) in Stanway to include a superstore site for Sainsbury's. Again secrecy over the negotiations was criticized by opposition councillors. The Borough Planner, though, saw this as an opportunity to resolve the traffic problems at Lexden. In 1984, when the Planning Committee was considering the Sainsbury's proposal for the Stanway store, the need to expand the car park at Lexden was again raised to make this point quite clear. The 67 500 sq. ft superstore was given permission in 1985. Sainsbury's have now closed the Lexden store but it is also rumoured that they may take the opportunity to close their town centre store.

The history of Tesco's involvement in Colchester provides another example of local retail change. In the 1950s Tesco's (another major supermarket chain) opened a store in High Street. This was a purpose-built store, on the site of a 17th-century building which was demolished. With the advent of new retailing methods, more space was needed and Tesco's moved to Head Street. Again a historic building was the chosen site but this time the façade and upper floors were kept whole and the ground floor was gutted. This 14 000 sq. ft store opened in 1966. Within a decade Tesco's were seeking a two-acre superstore but instead, in 1977, they moved to purpose-built premises with its own car parking in central Colchester.

Only four years later the decision to take the superstore site in the large residential development at Highwoods was announced. By 1985 it was proposed that the store there be doubled in size from 53 000 sq.ft to 102 000 sq. ft gross. Planning permission was duly forthcoming and tree-felling began to clear the site. Within ten months came the decision to close the town centre store. Even at the Highwoods site, expansion by Tesco's has affected local shopping facilities. A small square had been intended with several small shops as well as the superstore. However, Tesco's expansion reduced the number of other shops to two, and the use of doors directly onto the car park has taken the pedestrian flow away from this potential local feature.

As the Sainsbury's case illustrated, out-of-town retailing has encroached on the industrial estates in the town. The Peartree Business Centre is another victim of this trend. It was developed as an estate of small units for local business on a well located site to the west of the town, almost opposite two retail warehouses. An admitted bungle in the Planning Committee in 1984 resulted in approval for retail sales from a unit on the park (this was in line with the officers' recommendation, but a similar application was refused permission at the same committee!). Given this precedent the committee felt it was difficult to refuse later applications, so that many of the units are now effectively retail outlets. This has caused local traffic complaints as well as undermining the employment potential of the site.

Residential development

These difficulties in controlling retail development both on a particular site and over a broader area are also evident in the case of residential development. In Colchester this issue is dominated by the fate of the Highwoods estate (Fig. 4.1). This is a large area of land to the north of the town which, as the name suggests, incorporated many acres of attractive and well established woodland.

Highwoods was first proposed as a major residential development area in the growth period of the 1960s and was shown as such on the *Development plan review* adopted by the County Council. Some 573 acres were allocated and the planned increase in population was 18–20 000. At about this time the housebuilders, W. C. French Ltd, began buying land in the area, no doubt along with other builders. In 1969 it was announced that a scheme was being drawn up jointly by local and central government. A start was expected in two years.

In the early 1970s a proposals and design guide was issued by the council and an unofficial consortium was formed consisting of W. C. French Ltd and three other local builders. Negotiations ensued concerning issues such as the location of roads through the estate and the fate of an historic farmhouse on the edge of the site (which has since become the local pub for the estate). A go-ahead in principle was given by the town development committee in 1973. Further delays hit the proposal when central government temporarily withdrew funds for council road improvements, and more changes in road layout were made.

Then an apparently dramatic *volte face* occurred when, following the Local Government Act 1972, the new council took over from the old one. Previously, the Borough Council had favoured the development as had the County Council, but the new council was opposed to it. It was clear that once the transfer of legal planning powers under the Act had occurred the local planning authority would refuse the application. This duly happened by 29 votes to 22. The developers appealed to central government and in 1975 there was a local public inquiry. The decision the following year gave the developers their outline planning permission.

This would appear to be a case of a change of planning policy from encouraging development to restricting it, and the subsequent overruling of the local authority by central government to give development permission. But this oversimplifies events at the local level, both within the council and the housebuilders' consortium.

It was obvious at an early stage that the new council was not so favourably disposed to the development proposal, for the new council had been elected in May 1973 and sat 'in shadow' for ten months before assuming office. The councillors were partly seeking to demonstrate their appearance on the local political scene in a new organization. Then, to some extent, change of opinion reflected the new political composition of the council, with more rural councillors keen to limit the expansion of Colchester. And, in part, the council wished this decision to be taken at the central level, without any reflection on local councillors.

In any case the *volte face* did not amount to an outright rejection of any development. The new council sought to contain the scale of the development. Negotiations between the council officials and the builders had continued for some time on this basis. By the time of the inquiry the main issue was the amenity and traffic impact of allowing the development, particularly on the southern slopes of the site nearest to and overlooked by the town.

The planners were in some difficulty in arguing the case for limited development as they had only recently been seeking to encourage the project. In addition, the County Council was firmly in favour of the proposal, recognizing that if Highwoods was not developed it was inevitable that land would be released elsewhere, probably in the rural areas south of the town. Thus it was no surprise when the housebuilders won the appeal. The Secretary of State amended the plans to provide an additional 47.5 acres of open space to the south of the scheme, and made some amendments to the layout. Most of these amendments had been agreed by the builders at or before the inquiry.

Within the builders' consortium changes were also occurring which affected interpretation of the events. Although an unofficial consortium of four developers had submitted the planning application, French Kier (formed by the merger of W. C. French Ltd and J. L. Kier & Co. Ltd in 1973) were rapidly becoming the major landowner. By mid-1975 they owned almost 400 acres of the site. They achieved this position partly through their own purchases and partly through buying out 'consortium' partners, Frank Parker.

French Kier at this time were not as concerned about developing all the site as might be assumed and were therefore quite willing to trade with the council. First, the southern part of the site was the most expensive to develop owing to infrastructure problems. Secondly, there were problems within the recently merged company. While French were predominantly housebuilders, Kier were mainly civil engineers. In the early 1970s Kier were engaged in a number of motorway contracts for the government on a fixed-price basis. The oil crisis of 1973/4 and subsequent cost inflation placed the company in severe financial difficulties with collapse being rumoured but staved off by government assistance.

At the same time the crisis in the property market saw rapidly rising and then falling housing land prices. While the motorway contracts were undoubtedly the main source of internal problems, housebuilding was not viewed very favourably at board level. Until the recent take-over of French Kier by housebuilders, Beazer, the company continued to favour civil engineering projects over any substantial and sustained investment in housebuilding. The housebuilding activities were therefore largely based on the development of an historic landbank.

So, at the time of the Highwoods appeal, French Kier were willing to consider limiting the area of development but they wished to do so by selling the undeveloped land to the Council. Before the inquiry they had offered to sell 188 acres to the Council for a reported £2.75 million, reducing the new population of the area to 12 000. To some extent the public local inquiry concerned not so much an argument about restricting or allowing development as about the price at which land was to be sold to the Council: at development value with the benefits of planning permission or at current use value. The result of the appeal suggested the former alternative.

Subsequent changes in land ownership gave the Council some additional power. Gough Cooper, a member of the unofficial consortium, sold their land to the Council. This land was to the north, adjoining land the Council already owned and, most importantly, controlling access onto the site. This gave the Council a trump card. In 1977 it agreed a land swap, involving some additional consideration, and at the end of 1978 some 167 acres on the southern slopes were exchanged for 70 acres to the north. The first detailed planning permission for housing was given in spring 1978.

The granting of permission for development on the site was thus a long-drawn-out process. It illustrates an attempt by a Council, previously committed to encouraging development, to curtail that development. It also shows the negotiative form such attempts can take and the role of central government in intervening in such attempts. In particular, though, it illustrates the importance of land ownership by the Council in achieving its aims, even if such land had to be bought at above current use value. In the Culver Precinct case, land ownership by the Council seemed to strengthen its existing predisposition towards the development. In the Highwoods case, land ownership became the principal means of influencing the development, given the predominantly market-led nature of planning outcomes.

As far as the details of development at Highwoods are concerned, French Kier have not found the planners unduly restrictive. The local plans acknowledge the influence of market forces on the phasing of development and on residential densities. As at Culver Precinct, demand conditions dictated a fairly high standard of design and finish. French Kier cite minor problems in the form of unwanted cycleways, an unnecessary underpass and disagreements over tree-felling. When an issue threatens marketability, as over a Council-desired footpath linking middle- and higher-income housing areas, and when time and cashflow allow, French Kier have pushed on to appeal. A success in such an appeal is considered to reinforce the influence of Circular 22/80 (DOE 1980) on development control practice and Circular 15/84 (DOE 1984b) on housing land policy in placing market criteria centrally in development control decision-making.

The trend planning style

Planning within Colchester is not a streamlined form of rubber-stamping of development proposals. Rather, it seeks to combine the encouragement of private-sector development with a particular type of planning control. This example of trend planning in practice provides an opportunity to examine the institutional arrangements, politics and decision-making, and conflicts and tensions surrounding this planning style.

Institutional arrangements

As with other district councils, Colchester Borough Council has to operate the statutory planning system within a network of local authorities. It was shown in the case of Cambridge that this raises the potential for conflict between the different authorities, even though the relevant county and districts may be united on the general thrust of local planning. Conflicts in Essex have generally focused on the imposition, as the lower-tier authority sees it, of land allocations from above. This echoes the situation in some other authorities (Healey 1983, p. 49).

Since the Abercrombie plan of 1944, Essex has been regarded as a potential overspill area for population from Greater London. This theme was pursued in the 1964 *South East study* and the 1967 *Strategic plan for the South East* (SPSE). Major land allocations resulted, including at the new town of Basildon and substantial growth at Chelmsford and Colchester. The County Council, however, resisted some of this growth, particularly in the south of Essex, by extending the Green Belt in the 1964 *Development plan review* and the first structure plan (Elson 1986). The resulting conflict between central and local government was fought out at many planning appeals and at the 1980 Examination in Public (EIP).

By 1980, central government had come to see expansion in Essex as less urgent, given the downward revision of population forecasts after the 1960s, and the need to complement inner-city policy, particularly in the London Docklands. But its plans for the area were still more expansionary than those of the County Council. In the structure plan the County Council was seeking to limit any future development to existing commitments, i.e. land already allocated in the previous development plan. The result was the amendment of the structure plan by central government to allow for an extra 5200 dwellings.

Since then, the economic recession and the designation of Stansted as the third London airport have encouraged the County Council to move further towards central government's viewpoint on development in Essex. Nevertheless, it retains a strong commitment to the Green Belt and other ways of conserving the rural environment. As is common in the rhetoric of structure planning, the 1982 county strategy therefore contains two, potentially conflicting, themes: encouraging development and protecting the environment.

While the county has, until recently, been resisting the growth pressed on it by central government, Colchester Borough Council has been in conflict with the County Council over attempts to define its development allocation too precisely. The 1980 EIP saw the Borough Council, along with other district councils, criticizing the structure plan allocations for being inflexible and removing discretion from the district level. Larger allocations were sought by several district councils to enhance their role in local planning. Colchester Borough Council did not seek extra allocations but criticized the phasing policy that the County Council sought to impose (this was eventually deleted by the Secretary of State).

Conflicts between tiers of government have, therefore, taken two forms: central government and the County Council disagreeing over the share of regional growth to be accommodated; and the district councils and the County Council clashing over the power and responsibility for detailed land allocations.

Politics and decision-making

The generally responsive attitude of the Borough Planning Committee and officials to development is undoubtedly influenced by the long-term dominance of the Conservative Party. Since the new borough was formed in 1974, the Conservatives have formed the largest group, and from 1976 to 1986 had an overall majority. They have taken the lead in promoting Colchester as a location for private development. Opposition members, whether Labour or Alliance, have tended to take a more anti-developer, pro-conservation line. Yet this is tempered by a general acceptance, at least among Alliance councillors, that local planning outcomes must rely on the private sector. Therefore the fact that, with an increased Alliance presence, there is now a hung council does not appear to be substantially altering the council's attitude towards private development, which is still seen as leading urban change. Indeed, as in Cambridge, planning is not regarded as a very party-political issue. For this reason, the hung council is unlikely to generate more overt party-political conflict than under the Conservative-ruled council.

However, potential conflicts between councillors are only one aspect of planning decision-making. There is also scope for conflicts between planners and councillors, i.e. within the élite of key decision-makers. This is evident in relation to particular important development proposals, such as the Culver Precinct project where key individuals, such as the Town Clerk, the Borough Planner and the chairpersons of various committees, sought to influence the planning outcome.

It is also evident in the more day-to-day decision-making on planning applications. The planning committee of 13 councillors here has to deal with the influence of a strong planning officer. The Senior

Development Control Officer separates the 60 or so planning applications coming before the committee into two agendas. The 'white' agenda is discussed in full while the 'green' agenda, approximately half the applications, will be voted through collectively, unless any councillor wishes to raise a particular case. It is claimed that the wishes of the committee are usually anticipated to the extent that few 'green' agenda items are separately discussed.

The strong concern of the committee with aesthetic and architectural matters also echoes the Senior Development Control Officer's approach. However, it could not be said that, without his personal influence, planning decision-making would be less concerned with these aspects. For the concern with localized, environmental and primarily visual impacts of development characterizes local conservationist movements, and may have some of its roots in the pressure they place (or have placed) on the council.

In the past, the generally pro-development attitude of the council has created conflicts with local conservation groups, such as the Civic Society. As a result the Conservation Areas Advisory Group was set up. This comprises representatives of the Colchester and District Federation of Amenity Societies, the Civic Society, the Essex Archaeological Congress, and local architects and planners. It used to meet fortnightly to consider planning applications but now only convenes every two months. Items are not delayed in their passage to planning committee and thus many applications, for example for listed building consent, go through without their comments. The less active role of conservation groups in monitoring local development is not seen as a problem as the Senior Development Control Officer is considered to maintain adequate vigilance.

The legitimacy of local planning and the support of certain middle-class groups would be threatened if the local authority did not recognize a role for itself in ameliorating and influencing development. Therefore, in addition to amenity-conscious planning decisions, the council employs two enforcement officers, and enforcement notices can amount to 10% of the quantity of planning applications received in a year. Conservation Areas, including one covering the town centre, the listing of buildings and the scheduling of monuments also guide planning decisions in specific cases.

Therefore, even where local planning is oriented towards the market, professional influence remains important, as does the role of key councillors. In Colchester public pressure on environmental matters has been internalized by planning decision-makers, and individual applications remain subject to negotiations, mainly on design matters. There has been an attempt to follow current government thinking and place greater reliance on precedent, introducing elements of bureaucratic decision-making, but the streamlining effects are limited. Rather, there has been a tendency to maintain a generally open attitude to development pressure while controlling the details.

Decision-making in Colchester can therefore be characterized as non-strategic gatekeeping. Planners carry out the statutory planning procedures, but in a way that focuses on relatively minor issues while leaving major strategic issues to the judgment of the market. Professional concern is less focused on strategic planning and negotiation for planning gain than in regulative planning. There is generally less emphasis on spatial coordination and social planning skills. Attempts are made to expedite development control and a generally responsive approach to the private sector is adopted. Planners remain gatekeepers to the planning system but their attention is focused on aesthetic control and attempts to curtail the worst problems of anarchic development processes.

Conflicts and tensions

In taking on board conservationist concerns, many local planning documents and policy statements stress the need to balance the demands of the local economy and local environment in their rhetoric. But implementing such a policy would be a problem for a trend planning council. It wishes to encourage the private sector to develop in an area and yet, once attracted, the council seeks to control them. Too much control and development will go elsewhere; too little control and local amenities are threatened. Furthermore, the generally relaxed attitude to the level of development activity means that competition for particular sites is not very severe and hence developers are less willing to accede to detailed control by planners on any particular application.

Similarly, pressure from local conservationists leads planners to talk of managing growth or controlling development from the front, with a view to guiding it to particular locations or suggesting quantitative targets. But the generally pro-development attitude of the council undermines such a policy stance in practice. Rather, the planners' role takes the form of advertising areas to the private sector and then dealing with resulting planning applications, increasingly in a more bureaucratized manner by referring to precedent. Attempts to broaden this role come up against the lack of local political will effectively to control private developers' actions. Even where on occasion an attempt at regulation is attempted, central government intervention may undercut the negotiations, increased centralization reinforcing a generally laissez-faire local approach. These problems are evident in the various areas of development interest reviewed.

Culver Precinct Promoting development is never a costless process. Local advocates of trend planning would argue that in the case of the Culver Precinct project a compromise had been reached, and was satisfactory to both sides. They would point to the high quality of the development, the financial return to the ratepayers and the design concessions granted by the developer, such as retaining the library and certain listed buildings. Ultimately, though, their case rests on the achievement of the development being built at all. They would argue that account has to be taken of the strength of market pressure and the fact that town centre developments are being affected by the trend towards out-of-town retailing. In the early 1980s there was just enough pressure to enable development of this vacant site.

Critics would see many other aims being sacrificed to this one goal of achieving development by the private sector. The town square and high quality of finish are not seen as concessions, as they would probably feature in the blueprint for commercial reasons. On issues where profitability was threatened by local demands, as in the retention of the library as a central civic focus, the developer's wishes were followed. Even on the financial returns critics can query the final sums and point to the costs which the council has incurred. Up to late 1982 this amounted to about £1.25 million to buy the old library from Essex County Council, to undertake sewerage works, and as an advance payment to consultants (who eventually were to cost another £200 000 for their advice and liaison with the developers) (*Colchester Evening Gazette* 14 November 1982).

When Culver Precinct is complete, the prospects are that it will let easily and at rents almost twice those currently asked in the town's prime locations. However, it will divert demand from other parts of the town, and areas to the east of the newer precincts will decline. A department store in this area has already contracted. The next planning problem will then be to attract private investment to regenerate these areas. In retailing, in particular, letting the private developers lead urban change can result in economic activity shifting around a town, leaving the planners to follow in its wake and deal with the abandoned or declining sites. For even where the market is leading urban change, responsibility for dealing with adverse impacts does not lie with market actors, but rather remains to a large extent with the planning system.

Given these caveats as to the project's outcomes, what other benefits has the planning process achieved beyond facilitating the development itself and influencing its design? Another locally contentious issue is the extent to which planning has protected the ancient and historic aspects of the Colchester environment. Colchester was a Roman fortress and then a Roman town, and extensive remains of both periods exist. The Culver Precinct development lies within the boundaries of the fortress and thus in a very important archaeological area. Protection of remains is achieved through attaching a standard condition to planning permissions. This states that the council's agent should be given 'reasonable' facilities for access during the development to record items of archaeological importance. It also ensures that a timetable for such work is agreed before development commences. Such a condition was attached to the Culver Precinct permission and this allowed access to the Colchester Archaeological Trust. However, the limitations of such a procedure are clear (Joyce 1986).

In retrospect, the Trust estimated that to record and excavate the site adequately would have taken six months and £0.5 million. Back in 1979, the Trust had sent an implications report to the two shortlisted developers and asked for a contribution to the cost of works. GRE, the rejected developer, had been willing to meet the estimated cost of the excavation as part of the financial package. Carroll preferred a straight financial payment to the council and thus was able to offer more. In the event the money raised for the works, about £160 000, came almost exclusively from the council and the Commission for Historic Buildings and Ancient Monuments in equal shares. However, delays in the project and uncertainties in the demolition programme created problems for the Trust in keeping together a skilled team. By the middle of 1984 they were running out of funds. Carroll was approached and, in return for a renegotiation of the agreed timetable for access, gave £10 000. These difficulties were countered to some extent by the helpful attitude of the contractors, Balfour Beatty, who provided machinery and accommodation.

Nevertheless, the end result was that while some areas received the full attention they deserved, others were surveyed in a matter of days and some areas were not surveyed at all. Valuable items were left or destroyed. It is inevitable that such development destroys and covers over remains, but it also creates the opportunity to record the town's archaeological legacy. This is the role of 'rescue archaeology', not to protect or preserve. To achieve this, funds, time and access are essential. The Culver Street project, by virtue of the underground servicing (itself adopted for environmental reasons), destroyed a large number of remains. For example, in order to construct the service tunnel, Colchester's Roman wall, a scheduled ancient monument, had to be breached and a listed building demolished. The task of recording was a large one. Unfortunately, the verdict on the project has been 'insufficient funds, inadequate access and an enormous archaeological loss' (Crummie 1987, p. 245).

Out-of-town retailing The shifts in retail locations in Colchester, particularly the trend to decentralization, have greatly affected the built form and altered the focus of economic activity in and around the town. There are consequent effects of traffic movement and local amenities. Former premises can pose a problem for the council if the market does not readily provide a new user and hence the local environment is threatened. The council has therefore attempted to effect some control over the shifting retailers, even though this comes into conflict with its generally pro-development attitude. However, faced with local resistance to development pressures, central government has stepped in and upheld certain key appeals by retail developers. The council, mindful of the importance of precedent, has then felt impelled to grant other planning permissions. It could be argued that the imposition of development permission by central government lifts the responsibility from local shoulders, fitting in with the council's pro-development policy but sidestepping the legitimation problems of market-led planning.

The resulting dominance of market trends in retailing threatens not only the ability to provide local and accessible shopping facilities but also local employment opportunities, since several of the out-of-town stores are located on industrial estates. In 1978 the Chief Officer of the council was already admitting that:

...a possible criticism of existing Colchester Council industrial estates is that too many warehouses and shops have been allowed on the land. (*Essex County Standard* 29 December 1978)

The Conservative Chairman of the Planning Committee has stressed the need for flexible planning decisions which respond to the demands of business, but other councillors have questioned the impact of such a policy on employment. Profitable development in the area is predominantly retail warehousing, which has a very low employment:floorspace ratio. Profitable development can also result in relatively frequent relocation of business. In some cases the old premises can be taken up by a new user, but in many cases they become run-down. They may even stand empty. Abandoned retail warehouses on industrial sites are a potential problem in Colchester as their occupiers move to more attractive locations.

The latest council policy to deal with these problems is to identify commercial areas within industrial estates and guide rather than resist out-of-town retailing. But this has also been undermined by appeal decisions by central government. Recently, Texas Homecare were given planning permission on appeal for a site across the road from a council-designated commercial area (a permission which will result in the closure of their premises on the Severalls Park Industrial Estate). Following advice from professional consultants, the Council are considering allocating a 'retail park' for out-of-town shopping. Various sites are being assessed, but the success of this policy will be undermined if central government support for developer-selected sites continues. There is already a private-sector proposal for a suburban shopping complex in an industrial area, which pre-empts the Council's policy.

Highwoods   Residential development in the Highwoods area has been proceeding successfully, but the pace of development has been dependent on private-sector influences, rather than any council phasing programme. Some local-authority sheltered housing and some housing association developments were built to the north of the site, but the majority remained under the control of French Kier. They did develop some of this land themselves but, given the internal priority in the company for other construction activities, the pace was fairly slow. It was therefore decided to sell land north of the main spine road to other housebuilders such as Brosely and Barratts (who then sold their interest to Tarmac). These builders have built higher-density, smaller dwellings, including so-called starter homes, while French Kier have continued to concentrate on more up-market housing. The market for these starter homes can easily become saturated and building rates to the north have therefore also been slow. Other plots have been sold off individually with the benefit of planning permission, but these have not been very popular.

This pattern of development on the site creates problems for local planners. They have been charged by central government to ensure that sufficient land is available for housebuilding for the next two and five years. Where a large site accounts for a high proportion of the potential housebuilding land in an area, a slow-down in building rates means that the site will last for, say, 15 years instead of ten years. To maintain local housebuilding over the next two and five years, therefore, more land has to be allocated elsewhere locally.

This is the situation that Colchester Borough Council found itself in when preparing its local plans. Because of the rate of housebuilding at Highwoods, determined by intra-company and market demand factors, 77 extra acres of land and some additional plots had to be allocated elsewhere. Increased rates of building at Highwoods could theoretically lead to more land being developed than intended in the planning documents. For example, the take-over of French Kier by Beazer will probably speed up development. As at March 1986, only 45% of the site had been completed. The success of local planning in meeting structure plan targets depends largely on planners' ability to predict private-sector completions on this remaining land. This ability, in turn, rests largely on consultation with the builders involved.

Concluding comments   The reliance of this planning style on market outcomes generates significant tensions in implementation: the need to gauge the level of market pressure; the importance of land ownership in determining outcomes; and the difficulty of achieving any significant social goals such as the spatial coordination of land uses, the prevention of oversupply of developments and meeting local needs. Yet, at the same time, there is resistance from both professional concerns and local political pressures to withdrawal of planning from environmental protection. Without some evidence of planning activity which protects local residents' environment, the legitimacy of the local planning system is called into question.

Furthermore, the involvement of professional planners generates scope for the exercise of their discretion and judgment, which individual officers can zealously exercise and jealously guard from outside attacks. Developers can thus be faced with a local authority which overtly supports their role in local urban change and yet which, in practice, places obstacles in their path: design criteria, restrictions on development scale, and time-consuming consultation procedures. From the developer's point of view, the result is often an irritating concentration on planning negotiation over aesthetic, architectural and local environmental aspects of development. The political and professional pressures therefore limit the scope for a fully streamlined bureaucratic form of planning and result instead in the current non-strategic gate keeping mode. This suggests that the time is not yet ripe for any more substantial restructuring of the statutory planning system than that contained in the current, rather curtailed scheme for Simplified Planning Zones.

5

Popular planning:
Coin Street, London

Popular planning is planning by local communities in their own neighbourhoods. It involves both the formulation of planning proposals and their implementation by local community organizations. This rests on close collaboration between the community and the local planning authority, which has to be persuaded to adopt the popular plan as official policy. But the essence of popular planning is that local residents retain a high degree of direct control over the whole process.

For our case study of popular planning we have chosen to look at a small area of Central London known as Coin Street, which was the scene of a protracted fight between a major developer and a local community. In 1984 this struggle culminated in what has been described as 'one of the most extraordinary victories ever by a community group' (Cowan 1986, p. 6), when local residents gained control of the site and began to implement their own development scheme, since when many of the community's plans have been realized. In itself this makes Coin Street a classic case of popular planning, since few such plans have ever got this far – it may indeed come to be seen as 'the' classic case.

Coin Street and Waterloo

The area known as Coin Street is situated on the South Bank of the Thames in London, near the National Theatre (Figure 5.1). It consists of a string of sites, some 13 acres in area, lying mainly between Upper Ground and Stamford Street, which stretch from Waterloo Road through to the Thames at Stamford Wharf, with its famous 'OXO' tower. Like much of the South Bank, it has long remained

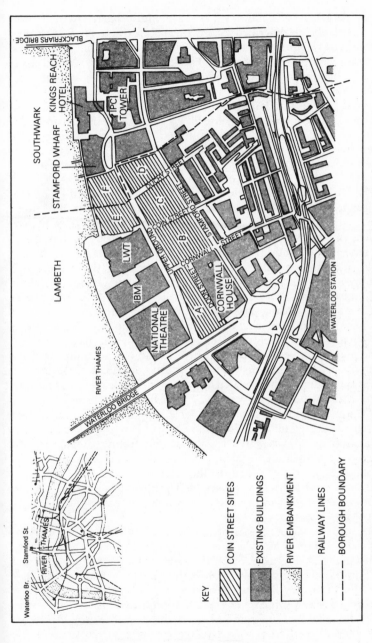

Figure 5.1 Map of the Coin Street area, showing places referred to in the case study

on the periphery of London's major land and property markets and can justifiably be described as a marginal area in economic terms. Before redevelopment began in 1986, most of the Coin Street area had been vacant for many years. The few remaining buildings were largely abandoned and the open land was used for temporary car parks. About half of the area was owned by the GLC, having been acquired by the London County Council (LCC) in 1953. Most of the remainder, including Stamford Wharf and the Eldorado Cold Store, was owned by the Vestey family through various companies, either freehold or on LCC/GLC leases. When this story began Coin Street represented one of the largest remaining undeveloped areas in Central London.

The Coin Street sites straddle the boundary between the London Boroughs of Southwark and Lambeth, falling mostly in the latter. The area forms part of the neighbourhood known as Waterloo which, like several other neighbourhoods in Central London, grew up around the station, completed in 1848. The main residential areas of Waterloo lie to the east and south of the station and comprise tenanted estates of the former GLC, the London Borough of Lambeth, the Peabody Trust and the Church Commissioners. Much of this housing was built on redevelopment sites between the wars, so little of the 19-century stock remains. Where there are Victorian and Georgian terraced houses these have attracted middle-class owner-occupiers. There is some local industry, particularly printing and distribution, mostly in small firms located under the arches of the elevated Waterloo and Charing Cross railway lines. Most residents of Waterloo work locally or in nearby parts of Central London and there are local shopping centres at Lower Marsh and The Cut. The Waterloo District Plan (London Borough of Lambeth 1977) remarks on the strong feeling of community among the remaining local population, down by half since 1961 to about 5000 in 1981. It is in the main a low-income, working-class community, with relatively high proportions of unskilled and semi-skilled workers and elderly households, not untypical of many inner-city areas today.

As well as having its local community, Waterloo is a part of Central London. It includes major office complexes, such as County Hall and the Shell Centre, St Thomas's Hospital, Lambeth Palace and the South Bank arts complex – the Festival Hall, the Hayward Gallery and the National Theatre. Waterloo Station itself is a dominant feature, covering some seven acres. Consequently, most people who work in and visit Waterloo come from outside the area. Although this gives Waterloo its attractive metropolitan character, it is the tension between the needs of local residents and the demands of outside interests which underlies the main planning conflicts in its recent history.

The initial conflict

The background to the popular plan for Coin Street can be found in a basic conflict over the future of Waterloo which came to a head in the 1970s. The conflict was between a future as part of the commercial expansion of Central London, through the speculative development of offices and hotels, and a future for the local community in the form of social rented housing and local employment and amenities. Before the 1970s, attempts to encourage commercial development in the area had been largely unsuccessful. In 1955 the South Bank was designated a Comprehensive Development Area (CDA), covering all of the Waterloo district between the railways and the Thames, and including about a quarter of the area's housing south of Stamford Street. Although the CDA was zoned for 'central area' uses, Waterloo was hardly touched by the office building boom of the late 1950s. London's Initial Development Plan of 1962 zoned the area for 'West End' uses, but again little new development occurred. It was not until the Greater London Development Plan of 1969 identified the South Bank as one of several 'preferred locations' for offices that any interest in redevelopment was stimulated.

The early 1970s saw the first commercial developments around Coin Street, but even in the midst of London's second major office boom, very little of this was speculative. The King's Reach hotel was built as a speculative venture but never completed, because its intended operator went into receivership (although the building was later converted into offices). However, there was further speculative interest in some of the Coin Street sites, and in 1971 the Heron Corporation was granted planning permission for an hotel on the site behind the National Theatre.

The same period saw the first organized responses from the local community. It appears that what first stirred the residents of Waterloo was a proposal to extend the Imperial War Museum into the adjacent public park. A welfare rights stall in the local market had identified various problems in the area, including a shortage of open space and play facilities. The extension to the museum would have taken up some of the existing open space, and so a campaign was launched to oppose it. This issue became a focus for community action and led to the formation in 1972 of the Waterloo Community Development Group (WCDG). The success of the campaign in stopping the museum extension inspired the WCDG to embark on the major step of developing a planning strategy for the Waterloo area.

Preparing a local plan

Commencing in 1973, the WCDG organized a series of public meetings and invited councillors and planners from the London Borough of Lambeth. The meetings discussed a wide range of local issues, including the changing types of shops related to new office developments, the closure of schools as the resident population declined and aged, and the shortage of low-cost housing. Housing seemed to hold the key, since it was needed to bring families back to the area and thus regenerate the demand for schools and shops, and Coin Street offered some obvious sites.

The GLC, which was the planning authority for the South Bank CDA, came under Labour control in 1973, and a similar series of meetings was held with GLC councillors and planning officers. As a result of the public meetings, both Lambeth and the GLC prepared independent reports on planning options for their respective areas of responsibility – Waterloo and the wider South Bank area. The reports offered a choice between private-sector office development with negotiated planning gains, public-sector housing development for local needs, or combinations of the two. In Lambeth the planners, like their colleagues in neighbouring Southwark, tended to favour office development in riverside areas, but the public's preference was for housing. Lambeth's politicians, unlike their Southwark counterparts, accepted this for most of Waterloo and it became the basis of the Borough's *Waterloo draft planning strategy*, adopted by the council in 1975 (London Borough of Lambeth 1975). The GLC adopted a similar policy in 1976, and this was published as *The future of the South Bank* (GLC 1976).

Lambeth was further persuaded by the WCDG to prepare a statutory local plan for Waterloo on the basis of the *Draft planning strategy*, and in 1977 the *Waterloo district plan* became the first local plan to be officially adopted in London (London Borough of Lambeth 1977). As a result Lambeth regained official responsibility for the Coin Street area from the GLC. Although not quite a popular plan, the *Waterloo district plan* 'bore the stamp of strong local approval and virtually no dissent' (Self 1979), following widespread public consultation. It included a policy of severe restraint on further office development and earmarked most of the Coin Street sites for housing and a public park.

The mid-1970s, the period during which these planning strategies were being prepared, was a time of retrenchment for the property development industry in London. The oil crisis of 1973/4, with its dramatic effects on interest rates and inflation, resulted in the virtual collapse of the speculative property market and the failure of several smaller banks (Rees & Lambert 1985). At Coin Street, the fate of the King's Reach hotel was only typical of other Central London developments, including the notorious Centrepoint office block, which remained unoccupied for years. Further interest in commercial development at Coin Street therefore subsided. The GLC broke off negotiations with the Heron Corporation and pressed ahead with the design of a housing scheme. In February 1977, the GLC gave scheme approval for some 200 dwellings on the two available sites in its ownership, the first stage of a plan to develop the whole Coin Street area for housing and open space.

The first phase of the Coin Street story underlines its relatively marginal position in the London property market. Commercial land uses were barely established on this part of the South Bank, with the exception of a few purpose-built complexes such as the London Weekend Television building and the International Publishing Corporation tower at King's Reach. Speculative property development had been tried on a small scale and had largely failed, and so remaining speculative interest appeared to have died away. Without very much effort, the field seemed open for the GLC to fulfil the objectives of the local plan and build housing for rent on some low-value redevelopment sites which it already owned. And this might well have been the end of the story: as *Coin Street News* put it:

> People would by now be living on Coin Street again had a new Tory GLC administration not axed the housing scheme and backed plans for a massive hotel and office project put forward by the Heron Corporation and Lord Vestey's Commercial Properties. (September 1984)

Property developers and popular planners

The election of a Conservative administration to the GLC in May 1977, under the leadership of Horace Cutler, heralded a new phase of property speculation at Coin Street, as in other parts of Central London. The respite of the mid-1970s had seen steady progress towards community goals – preparation of the anti-office local plan and the GLC housing scheme. But after toying with the idea of housing for sale, the new politicians at the GLC scrapped the housing scheme and, by expressing support for 'appropriate mixed developments' at Coin Street (Sudjic & Wood 1981), effectively declared their intention, as the major landowners, of ignoring the *Waterloo district plan*. This prompted Harry Dobin, a director of Heron, to declare: 'With the change of control at the GLC we thought we would get our plans out and dust them off' (*Tribune* 1 June 1979).

Political support for commercial development at Coin Street came not only from the predictable quarter of the Conservative GLC but from the less predictable minority Labour government. As Secretary of State for the Environment, Peter Shore contrived in August 1978 both to confirm the statutory status of the *Waterloo district plan* with its pro-housing, anti-office policies; and simultaneously to grant speculative office development permits to the Heron Corporation and the Vestey company, Commercial Properties, for over a million square feet of offices and a skyscraper hotel on the Coin Street sites. Since this positively invited planning applications contrary to Lambeth's declared policies, in the view of one commentator Shore had 'sold the pass' on the local community (Self 1979).

The community, however, was not standing still. In 1976, the large number of community groups in the area, including the WCDG, had formed an umbrella organization, the Association of Waterloo Groups (AWG), which was recognized by Lambeth as a neighbourhood council. The election of the Tory GLC prompted the formation of an active campaign group, the Coin Street Action Group (CSAG), to oppose the hotel/office proposals and promote housing and open space. Lambeth took over a version of the GLC scheme for Coin Street, consisting of 251, mainly low-rise, dwellings, to which it added schemes for the other two sites within the Borough boundary, and applied for a compulsory purchase order to acquire the sites from the GLC. The CSAG, however, was not satisfied that this scheme met the wishes of the local community and decided to prepare its own scheme for 360 low-rise dwellings, a riverside walk and park, shops and other facilities for all eight Coin Street sites, including those in Southwark.

In the confusion of competing and conflicting development proposals now seeking planning permission at Coin Street, in October 1978 the Secretary of State called-in all the applications for his consideration at a public inquiry. Even between this announcement and the start of the inquiry further proposals came forward. These included the community scheme, just mentioned; Heron's plans for an even taller skyscraper hotel – at 458 ft, potentially the tallest in Europe; London Weekend Television's application to extend its existing premises; and a third major mixed development proposal hurriedly tabled by a newcomer, Greycoat London Estates Limited (Greycoats). All of these applications were called-in for consideration at the inquiry, which opened on 22 May 1979.

The popular plan for Coin Street emerged out of a complex sequence of events over the next few years. The first public inquiry extended over 64 days and concluded in November 1979. Described by *The Times* (10 September 1984) as 'one of the longest, costliest and most important and confused planning inquiries ever held in Britain', perhaps its main achievement was to narrow the field and sharpen the conflicts. On the developers' side, Greycoats came out much the strongest contender. During the course of the inquiry, Greycoats submitted a revised scheme for the whole Coin Street area, designed by the international architect Richard Rogers. When it also acquired the freehold of the Boots factory and other leaseholds for around £2 million, Heron pulled out, leaving its partner, Commercial Properties, 'rather high and dry' (Milne 1979).

The community's development scheme for Coin Street was prepared by the CSAG. The Action Group worked by dividing its tasks among a large number of subgroups and calling on whatever sources of professional and technical help it could muster. These included the architect of the original GLC housing scheme; a worker in a local housing co-operative; lawyers attached to local law projects; a planner in Southwark; Shelter Housing Aid Centre; the Society for Co-operative Dwellings; and many other individuals. Publicity and public relations were central to their strategy: the Action Group produced a four-weekly bulletin and an occasional newspaper (*Coin Street News*), issued press releases, and organized exhibitions, a tape-slide show, street theatre and social events.

The community case was presented at the inquiry in a number of different ways. The formal planning application was submitted by the AWG, represented by a lawyer. Formal presentations of evidence in support of the community scheme were therefore made under the auspices of the AWG (hence it was generally known as 'the AWG scheme'). In making its case, the AWG was able to draw on a wide range of professionals and experts, including many of those who had helped in the preparation of the scheme. It also presented a unique analysis of supply and demand in the office market in Central London, commissioned from a planning consultant, in order to challenge the office location policy of the GLDP and to demonstrate that no more offices were needed. In parallel with the official proceedings the CSAG ran an action campaign, including a petition, publicity and demonstrations. Three or four people from the community groups also attended the inquiry full-time.

In July 1980 all the applications were refused by the new Conservative Environment Secretary, Michael Heseltine, who described the office proposals as 'massive and over-dominant', while criticizing the housing proposals because they 'failed to exploit the employment potential of the sites' (*Journal of Planning and Environment Law* 1983). Instead, he called for a mixed development which would combine housing and employment. Heseltine appeared to be defining a new planning policy for the area which incorporated elements from both the GLDP and the Waterloo District Plan but deviated significantly from both. In effect, he had thrown out a challenge to each set of developers, Greycoats and the AWG, to come back with comprehensive schemes which met the revised criteria.

Greycoats responded by joining forces with jilted Commercial Properties, to form Greycoat Commercial Estates Limited (for brevity, we will continue to refer to this company as Greycoats). This consolidated the private landholdings in the area, giving the new company control over about half the sites through a mix of freeholds and leaseholds. A revised scheme was published in March 1980 and submitted for planning approval in December. It consisted of a string of cluster blocks of varying height, linked by a glazed pedestrian mall and connected to a new Thames footbridge. Described by the architect as 'an open-ended flexible infrastructure capable of fostering a wide range of local and metropolitan activities' (Richard Rogers & Partners 1981, p. 52), the concept was much praised in the architectural press, while others nicknamed it 'The Dinosaur' and 'The Berlin Wall'. It amounted to a million square feet of offices (slightly less than the earlier version), housing, shopping, light industrial workshops and other facilities, including public open space. Almost immediately it was called-in and a second public inquiry became inevitable.

The AWG's revised proposal for a mixed development, comprising 400 dwellings, managed workshops, shopping and other facilities, and public open space, appeared early in 1980. It goes without saying that while both schemes included apparently similar elements, they represented radically different approaches to the development. The Greycoats scheme was a purely commercial venture which offered some social amenities as a planning gain, and was based on conventional institutional sources of funding. The AWG scheme was a thoroughgoing community project, which would provide low-rent housing for local people in need, funded either by the local authorities or a co-operative housing association; the managed workshops were mainly for light industrial uses, and were intended to extend the range of employment opportunities in Inner London; the shops would include a supermarket to supplement existing local facilities. The only common feature was public open space on the waterfront, and even here the two developments would have been unlikely to appeal to the same groups of users. There was no compromise between such diametrically opposed types of development, and it looked as if a conflict was about to become a battle.

When the second Coin Street inquiry opened on 7 April 1981, it was indeed 'amid scenes reminiscent of the worst motorway inquiries of the 70s' (*Building Design* 10 April 1981). The protestors, mainly local residents, were incensed that the AWG scheme was not on the agenda, and that the inquiry should be starting before the May elections for the GLC, when a Labour victory was (correctly) anticipated. The result was an adjournment until June and the inclusion of the AWG scheme to be examined alongside Greycoats' proposals. After a further adjournment on a technicality had again postponed the start of the inquiry until September, it ran for 88 days and closed in March 1982.

In the interval between the publication of the revised development schemes and the much-delayed start of the inquiry, two events, both involving the GLC, significantly changed the balance of forces in the field. The first was a deal concluded between the outgoing Tory GLC and Greycoats, in the form of a conditional Agreement for Sale which gave Greycoats an option to acquire all the GLC's freehold interests at Coin Street on condition that it secured all necessary planning and other permissions within three years. Greycoats maintained that this controversial deal was purely a commercial decision, to give them sufficient basis on which to proceed with their development. The GLC imposed restrictive conditions on the deal, in an attempt to ensure that the site was in fact developed, but GLC officers were clearly unhappy about making this agreement prior to the granting of detailed planning permission (GLC 1981a). It is hard to avoid the inference that the land deal was a political manoeuvre designed to prevent the successor administration at the GLC from blocking Greycoats' plans.

Whatever interpretation is put on it, the land deal neatly anticipated the second important event, namely the Labour victory at the GLC elections just mentioned, and the new administration's immediate decision to back the AWG scheme for Coin Street. By July the GLC had published a statement of its new policy, *The future of the South Bank wider area* (GLC 1981b). This aimed 'to limit the expansion of Central London activities into the South Bank. Housing should be the major land use with other supporting activities, such as industry.' Office development was to be restricted to sites specified in approved local plans, such as the *Waterloo district plan*. The new administration at the GLC also provided more practical support for the AWG, which suddenly found its resources boosted by the full-time secondment of an architect and almost unlimited use of copying and printing facilities.

As these moves imply, one of the first priorities of the new administration at the GLC was to protect all the working-class communities in Central and Inner London from the blighting effects of commercial development pressures. From July 1981 the Council began to set out its Community Areas Policy. Building on the South Bank initiative, this policy aimed to resist commercial development and gentrification in the old neighbourhoods surrounding the City and the West End, and to promote rented housing, community facilities and local employment, drawing for funds on the GLC's development programme. The areas covered by the policy ranged from Hammersmith to Spitalfields, and from King's Cross to Battersea. The South Bank, including Coin Street, was therefore defined as a Community Area and selected for funding from 1982/3 (GLC 1985a).

The community victory

The second Coin Street inquiry ranged over much the same ground as the first one, with both the AWG and Greycoats claiming that their proposals conformed with statutory planning policies for the area and with Heseltine's demand for suitable mixed development. The Secretary of State's decision was announced in December, just before his departure for the Ministry of Defence, and granted outline planning permission to both Greycoats and the AWG. The decision letter explained that both schemes were acceptable as comprehensive, mixed developments. This seemingly even-handed decision was widely seen to favour Greycoats, since it appeared to raise the value of the land beyond what the GLC could reasonably pay for housing and industry. But, undaunted, the AWG, under the headline 'Full Speed Ahead!', boldly announced its intention 'to start construction on site towards the end of 1984' (*Coin Street News* April 1983).

Greycoats' three-year purchase option had just over a year to run and it still needed road closure agreements and permission to demolish the Stamford Wharf building, since 1983 in a declared Conservation Area. Meanwhile the GLC, the London Boroughs of Lambeth and Southwark and the AWG jointly went to the High Court in an attempt to have Greycoats' planning permission quashed. Their contention was that the Secretary of State had acted improperly, in particular by failing to consider the supply and demand for offices, the provisions of the statutory local plan and the policies of the local planning authorities, of which all three now backed the AWG. Rejecting these arguments, Mr Justice Stephen Brown ruled in July that:

> the issue was not a question of 'housing against offices'; it was a question of whether the application proposals achieved an acceptable balance of a mixture of uses set in an appropriate architectural context, in accordance with the Minister's stated policy. (*Journal of Planning and Environment Law* 1983, p. 797)

An appeal to the Court of Appeal in December was similarly dismissed and a petition to the House of Lords was rejected, leaving Greycoats' planning permission intact but with the deadline on its purchase option rapidly approaching.

February 1984 saw the inquiry into the road closures required by the Greycoats scheme, actively opposed by the AWG and all three planning authorities, along with some 400 other individuals and groups, including King's Reach Developments. But no sooner did the inquiry close than Greycoats made a dramatic move:

> With its option on the GLC-owned land about to expire, no funding or tenants for its wall of offices, and demoralised by the persistent opposition to its scheme, Greycoat Commercial Estates and associated companies finally admitted defeat and sold their land interests to the GLC on 29 March 1984. (*Coin Street News* September 1984)

Greycoats appear in the end to have endorsed the view proclaimed on a banner strung across Stamford Wharf, that this was 'A Community Victory'. The developers were defeated by the combination of an extraordinarily effective local campaign and the considerable muscle of the GLC. In addition to failing to obtain all the necessary permissions to force the GLC to sell the rest of the site, notably consent to demolish Stamford Wharf itself, Greycoats realized that it would face community opposition all the way. The CSAG had threatened to organize further action, even a union Green Ban, which could seriously hamper the development. Greycoats did consider holding on to the site and blocking the AWG scheme but decided instead to sell up and concentrate its efforts in other, less contentious parts of Central London.

In fact, the developer's position had always looked rather precarious. Greycoats was only prepared to start construction once the offices had been pre-let, ideally with one tenant for each of the eight linked blocks of the scheme. As events dragged on into 1984, it was observed that the developer 'still has no firm potential tenants and, even more critically, no sign of major sources of development investment' (Milne 1984). Greycoats' change of heart was probably not uninfluenced by the start on site of the St Martin's Group development at Hay's Wharf, and by the fact that the company had recently secured two other major development projects in Central London. It sold its interests in the Boots site, Stamford and Nelson's wharves, and other smaller sites (amounting in total to some 6.5 acres) for £2.7 million.

George Nicholson, chair of the GLC Planning Committee, summed up the sense of euphoria which now came over the local campaigners:

> This is a landmark. It's the culmination of a long and determined battle by local people. The development we shall now see on this important London site is the people's plan – planning for the people and by the people. (GLC 1985a, p. 12)

Implementing the popular plan

With the whole of Coin Street in GLC freehold ownership, the AWG found itself in the spring of 1984 on the brink of realizing its popular plan. Although the tables appeared to have turned quite suddenly in its favour, the AWG and the GLC had been working for some time on a contingency plan. In 1983 a Joint Advisory Committee was formed, consisting of representatives from the GLC, the Boroughs of Lambeth and Southwark, and the AWG, with the aim of progressing the outline planning permission granted to the community scheme. There was an initial disagreement over who should act as overall developer. The GLC proposed that it should have this role, bringing in the Boroughs under joint committee and financing arrangements. Lambeth, in spite of its serious conflict with central government over spending levels and ratecapping, proposed that it should buy the sites and manage the development itself. However, neither of these arrangements was satisfactory to the AWG. It did not regard GLC ownership of the sites as secure, given the authority's imminent demise, while Lambeth councillors were fighting the government over ratecapping, and in any case were known to be opposed to co-operatives and wanted to develop conventional council housing. Drawing on grass-roots support in the Labour Party, the AWG was able to block both of these plans and take on the development role itself.

In December 1983, with the withdrawal of Greycoats looking more and more likely, it began the process of setting up a non-profit limited company to purchase the sites from the GLC once it had control of all the freeholds. In order to achieve a site valuation which the community group could afford, the GLC imposed restrictive covenants on the freeholds, effectively limiting the use of the land to the AWG scheme. By this means it was able to sell all the freeholds at an agreed value of £750 000 to the new company, Coin Street Community Builders (CSCB), formed jointly by members of the AWG and the North Southwark Community Development Group, in June 1984. CSCB financed the purchase with the aid of two mortgages, one from the GLC and one from the Greater London Enterprise Board, the repayments being covered by temporary income from car parks and advertising hoardings. The ownership of the freeholds and the income they generated gave CSCB the advantage of independence. It was able to employ five full-time workers; a company secretary and officers responsible for finance, housing and social facilities, commercial development and administration. A sixth full-time worker, an information officer, was funded by a small grant from Lambeth and Southwark.

The local community now owned 13 acres of Central London and, true to its ambitious prediction, the AWG actually had its project on site before the end of 1984, as demolition of the Boots building began, shortly followed by demolition of the Eldorado Cold Store. The scheme fell into three distinct parts with different problems of implementation: the housing, the river wall and walk and other public open space, and the other land uses (industry, shopping and leisure). The intention was to develop and manage the housing through co-operative housing associations. To achieve this, the housing sites were initially sold to the Society for Co-operative Dwellings (SCD), at the nominal value of £1, which acted as development agent while CSCB set up new primary and secondary housing co-ops. A mortgage was raised from Lambeth and Southwark Boroughs to finance the first scheme on Site C (Fig. 5.1). A final design was prepared, granted detailed planning permission by Lambeth and scheme approval by the DoE, and the first houses commenced on site in June 1986. It consisted of three-storey, six-person houses for families, including two eight-person units, and mostly with gardens.

The detailed arrangements for the development were complicated but critical to the future of this controversial scheme. The freeholds of all the housing sites were transferred to a new secondary housing co-operative, called Axle, and the lease for the first scheme to a primary co-operative, Mulberry. Apart from conforming with CSCB's co-operative principles, this form of ownership and management carried added advantages. For one thing it was exempt from the 'right to buy' under the 1980 Housing Act. If the housing had been developed by a local authority or conventional housing association, tenants would have had the right to purchase their own houses or flats at a discount, so taking them out of social ownership and beyond the means of households in need. It was also calculated to minimize the risk of the government finding some way to intervene and force the sale of the sites for commercial development.

AWG's planning permission required the construction of a new river wall and extension to the riverside walk before any buildings could be occupied. The GLC undertook to do this, together with the development of Sites D and F1 as public open space (Fig. 5.1), at an estimated cost of £4.5 million (GLC 1983). This also commenced in June 1986 on the basis of £2 million of forward funding from the GLC, agreed with the government prior to the authority's demise in April of that year. The successor to the GLC, the London Residuary Body (LRB), was unable to evade this financial commitment and the new walkway was opened in the autumn of 1987.

The other elements in the outline planning permission were 126 000 sq. ft of light industrial workshops and 67 000 sq. ft of shopping and leisure facilities, including a restaurant and museum in the restored Stamford Wharf building. Various sources of funding were explored for the estimated £6.75 million construction costs, in the public and private sectors (GLC 1983). Although it was operated as a charitable trust, the wharf was costing money to maintain and generating no income, so it was selected for the second phase of the development. In 1986 CSCB invited proposals for the use of the lower floors of the wharf, to supplement the 75 flats planned for the upper floors. Offices and luxury flats were ruled out, and tenders were invited to include workshops and a museum. Out of some 85 proposals two were shortlisted, one a children's museum, similar to the Halifax 'Eureka' project, and the other a Museum of the Thames. Both proposals came with independent development finance. At the same time one of the later housing sites in the programme (site E, Fig. 5.1) was designated for a temporary crafts market and workshops, modelled on Camden Lock. This left the greatest challenge for the co-operative developers in the planned third phase of the development, the managed workshops on the site behind the National Theatre. The scheme which they envisaged had implications for rent levels, lettings policy and training provision which were unlikely to be acceptable to a conventional institutional investor. Possible sources of finance included large private companies such as Shell or BAT, which had funded small, start-up workshops elsewhere, and the Greater London Enterprise Board. But, whatever the problems, CSCB was confident of its ability to realize the project and looked in a strong position to do so.

Popular planning as a planning style

Coin Street stands as a classic example of popular planning in the 1980s. There have been other cases of successful community opposition to major development schemes and a handful of local plans prepared in full consultation with local residents. The *Covent Garden action area plan* (GLC 1977), approved in 1977, which was largely based on a document prepared by local community groups, was perhaps the first example of a popular plan, but since then the community has not played a major role in its implementation. The *People's plan for the Royal Docks* (Newham Docklands Forum 1983), although it was a full local plan drawn up by Newham residents, only really stood as a statement of opposition to the LDDC and the STOLport proposal. But at Coin Street community involvement has passed through all the stages, from opposition through consultation and active participation, to the implementation of large-scale development within the framework of a popular plan. The Coin Street case study therefore provides unique insights into the processes of popular planning, its strengths and weaknesses, and its conflicts and tensions.

Institutional arrangements

The characteristic organizational form of popular planning is the community forum. The Skeffington report of 1969 first advocated the setting up of community forums for consultation with local residents in the preparation of local plans. One of the first was created in Covent Garden in 1973 as a 'representative' body, with members elected from among local residents, workers and property owners (Christensen 1979). In 1974 the Docklands Forum was created as an 'umbrella' organization for local community and interest groups (see Ch. 6). More recently, Sheffield set up a number of forums for consultation on its city centre plan (Alty & Darke 1987).

Although they vary in their style and range of activities, community forums have played a major role in planning consultation, acting as a focal point for a number of community groups and bringing them into the planning process. However, as an institutional form the community forum has some limitations. It exists essentially as a focus of communication between, on one side, the diverse social groups which form the community and, on the other side, the local authority. As such, the forum tends to be trapped in a 'consultative' role, invited to respond to local authority proposals but not expected to have any of its own. In trying to be representative it is not well placed to make positive decisions and move into active campaigning and real participation. In Covent Garden, this led to a split between the Forum and the Action Group, with the latter breaking away to engage in a more active campaign of positive planning. The Docklands Forum, although it has become a more active body since the designation of the LDDC, has also had campaign groups, such as the Joint Docklands Action Group, form around it. Generally, Skeffington-type consultative groups have suffered the fate of incorporation into local authority procedures, unable to take an independent critical line.

Significantly, Coin Street did not start with a Skeffington-type forum, set up by the local authorities for formal consultation with 'the public'. The initiative for a forum appears to have come instead from within the community, which put pressure on the local authorities (principally Lambeth and the GLC) to engage in consultation. The North Lambeth Multi-Services Group first identified local needs and opposed the War Museum extension, leading in 1972 to the formation of the Waterloo Community Development Group. This group, which paralleled another in the adjacent borough of Southwark (North Southwark Community Development Group), then became the main 'forum' for consultation on planning policy. At that stage it seems to have adopted a role similar to that of other community forums, receiving and commenting on the local authorities' documents and proposals. This group, then, carried the process of popular planning through the stages of opposition and consultation.

The formation of the Association of Waterloo Groups in 1976 was a further significant step. The AWG was established as an umbrella organization, with some 32 affiliated groups including the WCDG. While it took over the role of consultative 'forum', the Coin Street Action Group was set up specifically to fight the new commercial development proposals then emerging. It is interesting that many of the same people were actively involved in WCDG, AWG and CSAG, but that the different groups were used for different purposes. The AWG generally took on the mantle of the formal or quasi-official community group. We have seen how it presented the community case at the public inquiries, through a lawyer, and submitted planning applications for community proposals. The CSAG, on the other hand, was the activist wing, staging demonstrations and publicity events. The separation of these two organizations helped to maintain both the legitimacy of the AWG, in its relations with local authorities and formal planning procedures, and the independent voice of the CSAG. This tactic helped to sustain the impetus and dynamism of the active participation stage of the popular plan, leading to the relatively successful outcome of the 1981 inquiry.

Almost immediately after the AWG scheme was granted planning permission, along with the Greycoats scheme, a new phase of popular planning stimulated a further realignment of the community groups and their relationship with the local authorities. Initially, implementation depended on a closer relationship with these authorities (Lambeth, Southwark and the GLC) which would be the main sources of initial funding as well as the statutory planning authorities for detailed planning permissions. The authorities formed a member-level Joint Advisory Committee (initially within the GLC but later transferred to Lambeth), including representatives of the AWG, 'to co-ordinate and progress the proposals for the Coin Street site'. The committee worked on contingency plans for implementing the AWG scheme and tried to resolve the question of who should have overall responsibility for the development. As we have seen, the AWG won this important political skirmish, with the result that Coin Street Community Builders took over the freeholds of the development sites. In its turn, CSCB helped to set up a new consultative body, the Coin Street Development Group, to involve the community in the detailed implementation of the scheme, and established a series of primary and secondary co-operatives to develop the housing sites.

This rather convoluted history of community organization in the Coin Street case study shows that it is almost impossible to generalize about the institutional form of popular planning. The community forum advocated by Skeffington was never wholly successful, except as a consensual consultative body. At Coin Street, community activists demonstrated a rather sophisticated understanding of the roles of different kinds of community groups, which could represent various degrees of formality and informality, participation and opposition, in changing circumstances. They were aware both of the need for a formal relationship with the local authorities and of the dangers of political incorporation, and adopted what might be described as a 'horses for courses' approach to organization. Popular planning may well depend on this kind of organizational flexibility, based on a formally recognized umbrella organization such as the AWG but able to diversify and reform into a range of more specialized groups at different stages in the process.

Politics and decision-making

The Coin Street case involved a large number of interest groups, each having different kinds and degrees of power and each pursuing different objectives for the development of the area. Decisions came out of a shifting pattern of alliances, with groups forming and dissolving, and with frequent changes of political leadership in the respective public authorities. This form of decision-making can be described as 'imperfect pluralism', since not every interest is equally organized or represented, and decisions tend to be unpredictable and pragmatic. The eventual outcome of the events at Coin Street was not only unpredicted but regularly dismissed as unachievable, even by sympathetic commentators. The case illustrates a rather confused struggle for power in a situation where no one group, in the public or the private sector, held the upper hand for very long.

The idea of pluralism, however 'imperfect', suggests a political process to which all interests have access and no one is systematically excluded. In the case of popular planning, the obvious question is just how 'popular' is it? The apparent degree of pluralism suggested by the large number of community organizations involved at various stages may be exaggerated. Since many of the same people regularly reappear in different roles in different groups, it would seem that the community interest was being articulated by a fairly small group of activists. In spite of the large number of organizations in the area, a social survey in 1974 reported that only 6% of a random sample of local residents attended tenants' or residents' associations and 7% attended community associations (London Borough of Lambeth 1977, p. 19).

The representativeness of those involved has been an issue for the AWG and its offshoots. In a briefing note for local councillors the CSCB commented that the management committee of the first housing co-operative, Mulberry:

> is composed of six men and six women. They broadly represent the social make-up of the local community: two are printers, two retired, two unemployed, one is a teacher, one a receptionist, one a docker, one an administrator, one a housing advisor and one works full-time at Coin Street. (Coin Street Community Builders 1986, p. 10)

It was also reported that positive action was being taken to recruit a black committee member. However, while the sex, race and class of community representatives are undeniably important for their credibility and legitimacy, it goes without saying that they are no guarantee of socially progressive attitudes. Rather, what stands out in the case study is the consistent efforts of the AWG and other groups to achieve both wide participation and popular control, for example in their insistence on developing the housing as mutual co-operatives. The representativeness of the community groups is ultimately reflected in their consistent aims and achievements, which were always to do with the needs of the mainly working-class residents of Waterloo and the surrounding area.

The local authorities, with wider constituencies to serve, never had the same single-minded commitment to meeting such local needs. Through the mid-1970s, Lambeth and the GLC under Labour control supported community goals and planned to build council housing on some of the Coin Street sites. Southwark remained in favour of office development on Thames-side sites until 1982 when a new council was elected that was more sympathetic to local communities. Under Conservative control from 1977 to 1981, the GLC actively promoted office development. But after 1981 it was the Labour GLC which became the principal ally of the local community, much more committed to their cause than even Lambeth. (It was only after most Lambeth councillors were disqualified from office and a new council elected in 1986 that the authority came to support the idea of housing co-operatives, for example, and then rather tentatively.) The eventual success of the popular plan for Coin Street was uniquely due to the support and intervention of Ken Livingstone's administration at the GLC. Its Community Areas Policy established the principle of defending local communities in Central and inner London against the threat of commercial development and gentrification. This policy was later incorporated in proposed alterations to the GLDP which were submitted to the Secretary of State but never approved (GLC 1984b). Nevertheless, it led to the funding of many small projects through the GLC's development programme, including some housing schemes. The Coin Street project received considerable assistance and effectively became the flagship of Community Areas Policy, a major rebuff to a large commercial developer and a demonstration of what could be achieved, apparently against all the odds. It was also, of course, one of the GLC's grandest swansongs.

Conflicts and tensions in popular planning

Although it might appear to be a consensual process within the community, popular planning also generates conflicts and tensions. Generally, the wider the involvement in decision-making, the more potentially conflicting needs will be identified. At Coin Street there seems to have been a remarkably consistent view within the local community of what was needed. When a few of the Lambeth sites were being considered in the early 1970s, the consensus was for housing, principally houses with gardens. When the idea of a larger scheme emerged during the public inquiries, open space, workshops and social amenities were added to the original housing proposal.

The community itself does not therefore seem to have been in conflict over what to do with Coin Street. But conflicting demands have arisen in the sense of who should benefit from the popular plan and who should control its implementation.

The key tension at Coin Street emerged in the relationship of the community organizations with the local planning authorities. At various times and with various authorities this was a straight conflict of directly opposed aims; for example, with the Tory GLC and to a slightly lesser extent with Southwark before 1982. But even where the community and the local authority appeared to share the same goals, tensions emerged. The first housing scheme at Coin Street was funded jointly by Lambeth and Southwark, out of their Housing Investment Programme allocations. Although 90% of these loans would be repaid on completion of the scheme, through a Housing Association Grant, Lambeth insisted on 100% nomination of the initial tenancies from its own waiting list. To some extent this was an issue of who should benefit, the residents of Waterloo who had fought for ten years or people from other parts of Lambeth who might be in objectively greater housing need. It has been suggested that racial tensions were also involved, which the National Front attempted to exploit; Waterloo is a mainly white area and Lambeth had a policy of allocating at least 30% of new housing to black people (*City Limits* 29 March – 4 April 1985). The CSCB conceded the principle of nomination for the first scheme, but in order to be able to set up a mutual co-operative among the new tenants it insisted on nominations being made six months in advance of occupation and a full co-operative training programme.

The fact that at Coin Street the community has become the developer puts it in a unique relation to the planning authorities, and yet it is a position which is not dissimilar to that of any commercial developer. On the one hand the community owns the land and has an outline planning permission, but on the other hand it still needs detailed permissions and, perhaps more significantly, financial support from the authorities. Where commercial developers might only need publicly provided infrastructure, CSCB needs more direct help in the form of housing loans and the provision of social facilities. Some of the Coin Street development will be independently financed, like a commercial development, but there will always be an element of dependence on the local authorities and therefore tension over policy decisions. This would seem to be an inevitable characteristic of popular planning.

A further tension can be seen in the Coin Street case which is also characteristic of popular planning generally, and that is the question of the longer-term future of the plan. National government policies have been stacked against popular planning since 1979, if not before. The increased emphasis on market criteria in development control decisions, the 'right to buy' social rented housing, and the abolition of the GLC all worked to the disadvantage of the AWG scheme. Highly conscious of this problem, the AWG sought to maximize its independence and therefore control over the implementation of the scheme, with remarkable success. It also stuck firmly to the principle of housing co-operatives, which fall outside the 'right to buy'. In fact, the future of Coin Street looks reasonably secure at the time of writing (1987). Interest in speculative office development has waned on the South Bank, with the construction of London Bridge City in North Southwark and the shift of attention to Docklands and Canary Wharf. For the time being, the pressure is off and CSCB is able to get on with the development.

6

Leverage planning:
the London Docklands
Development Corporation

Leverage planning is the use of public investment to stimulate a weak or flagging private market in land and property development. As we pointed out in Chapter 2, this is an established aspect of the British planning system, in forms ranging from house improvement grants to Assisted Areas. But only in the 1980s has leverage become one of the principal approaches to the regeneration of declining urban areas. From being a relatively minor part of planning, leverage has become a mainstream activity which looks set to expand its role even further. Already, leverage planning has generated new institutional forms and distinctive political features. Its implications for urban renewal, in terms of the renewal process, the pattern of redevelopment and the social groups likely to benefit, are becoming clear. In short, we can identify the principal features of leverage as a planning style.

For our case study of leverage planning we have chosen the London Docklands Development Corporation (LDDC) (Fig. 6.1). The LDDC was one of the first two urban development corporations (the other was on Merseyside). Its essential role has been the preparation and marketing of development sites, often involving major reclamation works and the provision of suitable infrastructure, thereby turning large areas of worthless and derelict land into viable propositions for speculative property developers. It is in this sense that the style of planning represented by the LDDC can be described as leverage. New private-sector investment is neither incidental to its activities, nor part of a wider publicly determined planning scheme, but the principal rationale of the LDDC's investments and a major source of its revenue.

To begin this case study, and to help to understand the nature of the LDDC and the controversy which surrounds it, we will briefly review the decline of London's Docklands and the initial attempts in the 1970s to draw up and carry out major plans for their renewal and redevelopment.

96

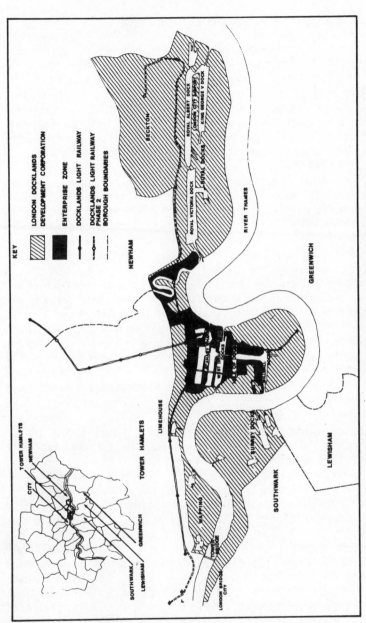

Figure 6.1 Map of the London Docklands, showing places referred to in the case study

KEY

London Docklands Development Corporation

Enterprise Zone

Docklands Light Railway

Docklands Light Railway Phase 2

Borough Boundaries

The decline of London's docks

London's enclosed docks – the upstream docks – were constructed during the 19th century, when London was the busiest port in the world. A variety of heavy industries grew up alongside the port, including shipbuilding, engineering and refining, together with shipping service industries. The East End also became the home for many obnoxious industries that were not tolerated elsewhere in London. Docklands in its heyday was therefore a major centre of industrial and commercial activity, employing large numbers of manual workers.

The collapse of the Docklands economy first became apparent in the 1960s, but its demise was deeply rooted. In a well documented account, Hardy (1983a) argues that laissez-faire competition for shipping trade led to an early overprovision of enclosed docks which ultimately was a major cause of their downfall. Even from the beginning, the dock companies struggled to remain profitable. They began to fail and merge as early as 1838, until only three companies remained to form the Port of London Authority (PLA) in 1909. Low profitability meant inadequate investment and eventually loss of business to more modern ports elsewhere in Britain and Europe. During this century the decline of the docks also reflected changes in the pattern of world trade, and changes in technology. London has lost most of the trade on which the Docklands grew up. The types of goods carried by sea are no longer those which sustained the industries of the East End, and London no longer functions as the centre of trade within an Empire. Technological changes have produced larger ships and metal containers, favouring deep-water terminals and ports with better road and rail connections. By the end of the 1960s the upstream docks were largely redundant. The first to close was the East India Dock in 1967; many of the others had followed by 1970 and the remaining docks, the West India/Millwall complex and the huge expanse of the Royals, had all closed by 1981. Direct employment in the docks fell from just under 23 000 in 1967 to a little over 7000 in 1979 (Newman & Mayo 1981, p. 535).

As the East End of London ceased to have a major role as a port, so it also declined as a centre for manufacturing industry. Decline has characterized Britain's older industrial areas since the mid-1960s. Newman & Mayo cite evidence that manufacturing employment in London fell by 34% between 1961 and 1974, half of this decline resulting from factory closures. The process of industrial restructuring continued during the 1970s, and with it came a relentless loss of jobs. There was a 27% decline in employment in the Docklands area between 1978 and 1981 alone (LDDC 1986a, p.

5). While some major factories remain, notably along the riverside south of the Royal Docks, industrial employment has continued to fall during the 1980s. As a consequence, unemployment in the Docklands rose to very high levels, reaching 18.6% in the LDDC area in 1981 (GLC 1984a, p. 21).

Like most other old industrial areas in Britain, the decline of the Docklands economy has been associated with a massive loss of population. People began to leave Docklands and the East End of London long before its economic problems became so overwhelming. The Docklands boroughs have been losing population since the 1930s but the outward movement accelerated in the 1950s. Better housing and employment opportunities in the overspill estates, the spec-built suburbs and the new towns attracted thousands from East London. By the late 1960s that movement had become an exodus, the five Docklands boroughs losing 10% of their population between 1966 and 1971. During the 1970s the 'pull' of opportunities was reinforced by the 'push' of decline, and several inner London boroughs lost over a fifth of their population in the decade 1971–81. It is estimated that the population of Docklands fell by 24% in this period, and when the LDDC area was defined in 1981 its recorded population was just 39 700 (GLC 1984a, p. 17).

Planning for Docklands in the 1970s

The need to plan for the renewal of the Docklands was recognized as soon as the first docks begans to close. In 1971 the Department of the Environment and the GLC commissioned a firm of consultants, R. Travers Morgan, to set up the London Docklands Study Team and prepare a range of comprehensive planning strategies. The Study Team's report (London Docklands Study Team 1973) set out five broad-brush strategies. Four of these were varieties of commercial development, including owner-occupied housing, luxury hotels, marinas, offices and shopping centres, with a new rapid-transit system. The fifth, described as 'East End Consolidated', was addressed more to local planning and employment needs but with 'all the appearance of a token plan to satisfy local traditionalists' (Hardy 1983a, p. 18). While acknowledging that political choices had to be made, the Study Team had in effect identified the potential for commercial investment in Docklands, in association with appropriate publicly provided infrastructure.

The Travers Morgan report received a hostile reception in Docklands itself. On one level it represented a solution imposed on Labour-controlled boroughs by a Conservative central government.

When control of the GLC was transferred to the Labour Party in 1973, the odds were stacked against it, and in 1974 Geoffrey Rippon, as Environment Secretary, withdrew the whole report (Hardy 1983a, p. 18). Economic circumstances had also changed dramatically in 1973–4 with the collapse of the property boom, and the future of all major developments was thrown in doubt. But behind these immediate circumstances, the London Docklands Study Team represented a particular conception of the planning process. It put forward large-scale proposals, drawn up without any consultation with the elected political bodies responsible for planning in the area nor, indeed, with local residents. Yet this was just the time when the planning system was beginning to accommodate demands for public consultation and participation in plan-making, through structure and local plans. The Study Team approach was therefore an unacceptable planning style (Ledgerwood 1985).

In looking for a way forward, Rippon set up the Docklands Joint Committee (DJC), made up of representatives of all the local authorities concerned with the area, together with the Port of London Authority and the TUC. Described by Hardy as 'a blue-print for appeasement, straight off the pluralist drawing board' (1983a, p. 18), this formula was the same as that which Rippon had applied to Covent Garden. The 'public' were represented through a Docklands Forum and involved in a wide range of public participation exercises, similar in format to those being developed in structure and local planning. In just two years the DJC produced the London Docklands Strategic Plan (LDSP), an advisory planning framework based on a single strategy agreed among the participating authorities.

Given its social and political pedigree, it is hardly surprising that the LDSP proposed almost the exact opposite of the Travers Morgan report. It argued for change – indeed, it proposed a doubling of Docklands' population – but for change which would consolidate the existing social and economic character of the area rather than radically transforming it. The plan was based on an assessment of the needs of the local population in the five Docklands Boroughs, Tower Hamlets, Newham, Southwark, Lewisham and Greenwich. It proposed to used the vast areas of land which were becoming available in Docklands to meet these needs, principally for housing (mainly local authority rented housing), industrial employment, social facilities and environmental improvements. It was a plan to make the East End of London a more pleasant place to live, for the benefit of the East Enders. Nowadays such a plan is easily dismissed as 'a working class Shangri-la' (*The Times* 19 April 1985) or 'a domestic paradise of waterside communities' (Pawley 1986). But as Ambrose (1986) has pointed out, in many respects the LDSP was a return to the principles of Abercrombie's County of London Plan of 1944, in which the Docklands were intended to retain their 'East End' land uses rather than develop higher-value 'West End' functions. It was a plan for the existing community, as they perceived their interests in the mid-1970s, and not a plan for some undefined community of anonymous newcomers.

Briefly, in the late 1970s, the LDSP began to be implemented, taking advantage of the Labour government's inner cities programme and other special funds. But it quickly ran into political, practical and financial problems and progress was painfully slow. The change to Conservative control at the GLC in 1977 put the five Labour boroughs at odds with their strategic planning authority. All the major programmes fell behind schedule, as public expenditure was cut after the 1976 financial crisis, and manufacturing industry in the inner-city areas entered a steep decline. Land reclamation and site assembly proved much more costly and complicated than envisaged, and the Docklands Joint Committee had no effective powers. The legacy of the LDSP from the period 1976–81 was the filling of some docks, notably the Surrey Docks (later criticized because waterside areas provide commercially attractive sites), the construction of some 2500 local authority dwellings, a few social and community facilities, and plans for road improvements which were broadly accepted by the LDDC. Given the context in which it was being implemented, and in particular the massive withdrawal of private investment and major cuts in public expenditure that characterized the fate of most inner-city areas in the late 1970s, this modest record is easily explained. But it has also been argued that a locally conceived and administered solution was doomed from the start, and that a plan such as the LDSP could only become a reality through radical reform of the planning and development process, including control over private investment (Newman & Mayo 1981). Be that as it may, the poor track record of the Docklands Joint Committee presented an easy target for criticism by the incoming Thatcher government in 1979.

The London Docklands Development Corporation

The LDDC began operating in Docklands in September 1981, just two years after its announcement by Michael Heseltine, the Secretary of State for the Environment who shaped the Conservative's policies on inner-city renewal. The legal basis for urban development corporations (UDCs) was created by the Local Government, Planning and Land Act 1980. Partly modelled on the New Town Development

101

Corporations of 1946, UDCs were designed to take over some responsibilities of the existing local authorities in their designated area, principally those for development control, and were given the power to buy and sell land. The boundaries of the proposed LDDC included an area of 5120 acres. This was a slightly smaller area than that covered by the Docklands Joint Committee since it excluded those areas which fell in the boroughs of Greenwich and Lewisham. The three boroughs affected by the proposal (Tower Hamlets, Newham and Southwark) objected to the designation and a hearing in the House of Lords ensued.

The House of Lords hearing revealed much of the thinking behind the new style of leverage planning which has now been extended to other inner-city areas. There were two key arguments advanced in favour, one concerning the nature of the problem and the other the nature of the solution. The problem, the dramatic decline of London's Docklands, was said to be of national as well as local significance. This reflected the proximity of the area to the City and Central London and the priority given to the capital by the government. It was also a matter of the scale of the problem in relation to the resources required. The solution, the government claimed, required a single-purpose authority to overcome the limitations of existing planning procedures and organizations. What this meant was that the government did not believe the DJC and the boroughs would be attracting private investment for the regeneration of Docklands. Substantial new public investment was needed to change the image of Docklands and to create conditions which would bring in the private investor, and this could only be channelled through a new body which would not spend it on 'local needs'.

The boroughs and local community groups challenged these arguments with the claim that the DJC was an 'efficient and responsive organisation' (Colenutt & Lowe 1981, p. 236) which had already achieved much, notably dock filling and site preparation. But on the key issue of private-sector investment the objectors' case was unclear. Southwark and Tower Hamlets claimed that they welcomed major private investment in industrial and commercial projects, such as the St Katharine Docks and the News International plant in Wapping, and the Hay's Wharf scheme in Bermondsey (now London Bridge City). However, the community groups argued that expensive owner-occupied housing would not benefit local residents, and Newham made a strong case for more council housing in its area. The LDSP had certainly required private investment to achieve its goals, but in practice it had not proved attractive to those parts of the private sector which were interested in investing in Docklands. The Lords Select Committee concluded that 'private investors will not put money into Docklands on any large scale unless they are encouraged by the presence of an environment attractive to them, including the availability of some private housing' (House of Lords 1981), and recommended in favour of the LDDC.

The LDDC was thus established as a new kind of planning instrument, with the task of regenerating the Docklands in a new way. It has three main characteristics which distinguish it from other styles of planning. Firstly, it is a distinctive type of organization, a quasi-governmental agency. As such it falls outside the control of locally elected politicians but is accountable to Parliament through the Secretary of State for the Environment. Secondly, the LDDC decided 'to use its resources primarily as a lever to attract private investment' (1986a, p. 3). These resources include not only finance but also exceptional powers to acquire and prepare development sites. With its general object defined as 'to secure the regeneration of its area', the LDDC has pursued the clear policy line that this can best be achieved by encouraging private-sector investment. In pursuit of this goal, its third main characteristic is a flexible approach to planning. The existing statutory plans for the Docklands are, in principle, unaffected by the presence of the LDDC, which has no plan-making powers. Instead, it purports to take account of these plans, along with the LDSP, in its development control decisions. However, few would argue that the statutory plans have acted as a constraint on development. In fact, the LDDC has defined its own flexible planning frameworks: as its former Chief Executive, Reginald Ward, candidly put it, 'it was necessary to be opportunist with regard to proposals from developers' (1986, p. 118).

The renewal of Docklands

We turn now to examine the effects of leverage planning in action, by reviewing the major activities of the LDDC and the development that has occurred in Docklands in the six years following its designation. By 1987 the LDDC had spent some £340 million and was receiving an annual grant-in-aid of £58 million. It was projecting total expenditure of around £1 billion up to 1991, with about half coming from government grants and the rest from the sale of treated and serviced land (LDDC 1987). More than three-quarters of the Corporation's expenditure was incurred on land acquisition, reclamation and treatment, including infrastructure works, with the remainder covering administration, consultancies, promotion and other costs. The major investments of the LDDC can be grouped under two broad headings, communications, and site acquisition and preparation, both representing major infrastructure development.

Improved communications for Docklands were seen as a key to its regeneration by the original Study Team and by the DJC, and this is perhaps the one issue on which there has been a large measure of agreement. A rapid-transit link with the City and London's underground system has always been part of this general proposal, and with the collapse of the plan for the Jubilee Line underground extension in the late 1970s, this eventually materialized as the Docklands Light Railway (DLR) (Fig. 6.1). The first phase of the DLR, built partly on the route of an old railway line, was opened in 1987 at a cost of £77 million (funded jointly by the LDDC and London Regional Transport). Eastward extensions are proposed to connect the Royal Docks and Beckton, at a further cost of some £230 million. In addition to the DLR, new roads have been an important infrastructure investment, since access to the docks was previously very restricted. Major road improvements are planned to both east–west and north–south routes, with the long-term aim of linking into the new East London River Crossing planned for the mid-1990s. There has also been an emphasis on telecommunications and cabling to service high-tech industry and offices.

Site preparation is the other major activity of the LDDC. Dock filling was completed early on at London Docks in Wapping, since when it was decided to retain all the remaining docks. Site works have ranged from the basic provision of sewers and access roads, as in the Surrey Docks, to major engineering works such as the construction of new river walls and the treatment of heavily polluted sites. All of these works were seen to be necessary to overcome negative land values and to make sites attractive to private investors. If land values continue to rise, it is possible that some of the future infrastructure will be provided by developers, for example at the Royal Docks.

Investment by the LDDC, substantial as it is, is dwarfed by that from the private sector. The Corporation claimed that by 1987, every £1 it had spent had brought in £8.72 from private investors. In the long run, it was anticipating over £6 billion in private investment, representing a potential 'leverage ratio' of over 12 : 1 (LDDC 1987). This leverage calculation is a rather misleading construct, since it only accounts for new public and private investment since 1981, and the public-sector element only includes net spending by the LDDC. Important factors omitted are infrastructure investment by the boroughs and the DJC prior to 1981; Urban Programme and other public-sector investment after 1981, such as housing and trunk roads, and the DLR, additional to the LDDC budget; rate and tax allowances in the Enterprise Zone; and continuing private-sector

104

disinvestment. As a broad indicator, the leverage ratio clearly points to the revival of an active land and property market in the past few years, which is the key object of this style of planning. In one sense, this still has to be judged with caution, since nearly all of the private investment is speculative, principally in offices and housing. While the housing is selling well to owner–occupiers, this is not yet the case with the offices: 'Still required are the votes of confidence of office users and the funding institutions who have so far shown little enthusiasm for Docklands' (Knight, Frank & Rutley 1987, p. 12). Nevertheless, the scale of regeneration now taking place shows that the intervention of the LDDC has made the area attractive and available to large-scale private investment.

A brief tour of Docklands will give some indication of what this money is being spent on and how the whole area is being transformed. Docklands falls into four geographical zones: to the north of the river, Wapping and Limehouse; the Isle of Dogs; and the Royal Docks and Beckton; south of the river, the Surrey Docks (Fig. 6.1). These are treated as separate planning zones by the LDDC, each with an Area Team, and they have begun to develop distinctive characters.

Wapping and Limehouse

At the western end of Docklands and closest to the City, Wapping was the first area to attract commercial investment. It stretches from the St Katharine Docks, by Tower Bridge, to Shadwell further downstream and includes the site of the filled London Docks. The St Katharine Docks were the first to be redeveloped. As soon as the docks closed in 1969 developers were queueing up for this plum site and the prize went to Taylor Woodrow. The St Katharine-by-the-Tower Hotel was opened in the early 1970s and the scheme now includes a marina, high-class shopping and, more recently, the World Trade Centre. Protest about this development in the early 1970s helped to form the Joint Docklands Action Group (JDAG), a constant critic of private-sector renewal. Today St Katharines is a major tourist and commercial success which symbolizes the wholesale transformation of the social and economic character of Docklands.

Wapping has other attractive features apart from St Katharines. There are several majestic wharf buildings on the riverfront, together with a fine group of Georgian houses at Wapping Pierhead, a conservation area. The oddly named High Street runs between tall warehouses and the area contains many listed original dock walls. These features, together with its location, have made Wapping a prime site for luxury housing at fabulous prices. Riverside wharves have been converted into flats with basement car parks, porterage and security, producing values of up to £275 per square foot (1987). The Barratts conversions at Gun Wharf were selling in early 1987 at prices from £95 000 for a studio apartment to over £285 000 for a penthouse. One of the latest conversions, New Crane Wharf, includes condominium-style leisure and recreation facilities on the lower floors. This type of residential market is closely linked with the 'Big Bang', the deregulation of the City of London in 1986. High salaries and a longer working day have led more City workers to seek 'pied-à-terre' accommodation near their offices (Knight, Frank & Rutley 1987).

Away from the river there are two other types of new housing. The 30-acre Western Dock site has been developed by speculative house builders with nearly 100 flats and townhouses, some of them alongside a newly built canal. In total, the new housing schemes are expected to double the area's population by 1990; in 1987 some 2000 new dwellings had been started, with a further 1700 in the pipeline.

The other type of housing development away from the river is the conversion of former local authority flats for owner-occupation. Most of Wapping's 3100 existing residents were tenants of Tower Hamlets, the GLC or charitable housing trusts. Two major schemes, the Waterlow Estate by Barratts and Riverside Mansions by Regalian, raised the controversial question of affordability by existing tenants. While some may have benefitted from windfall gains, others were displaced and most of the flats were lost from the council's stock. Both of these schemes were made particularly attractive by their proximity to DLR stations.

While it is becoming a predominantly residential area, Wapping has also seen a large influx of commercial development. One of the first to move in, before the establishment of the LDDC, was Rupert Murdoch's News International print works, 'Fortress Wapping'. With the eastward expansion of the City of London, following deregulation in 1986, offices have also come to Wapping, particularly smaller suites. The move of the London Commodity Exchange to the Royal Mint site near the St Katharine Docks (just outside the LDDC boundary) was seen as a 'stamp of approval' for Wapping as an office location. A large mixed commercial development at Tobacco Dock, which included shopping, leisure and workshop spaces, was expected to compare in its attractions with Covent Garden Market, but with twice the floorspace. With some sites changing hands at £3 million per acre (1986) Wapping's commercial future looks assured. Boasting that 'the area provides a striking example of the correlation between the Corporation's investment programme and the large scale private investment that has been attracted to bring forward regeneration' (LDDC 1986a, p. 49), the LDDC anticipated that the process would be completed by 1989.

Just to the east of Wapping, Limehouse Basin presents a rather different story. A mixed housing and commercial scheme for the dock basin ran into organized local opposition. People living and working there formed the Limehouse Development Group to put forward an alternative to Richard Seifert's £70 million scheme for Hunting Gate Homes, British Waterways' preferred developer. A public inquiry criticized the Seifert scheme for overdevelopment of the site, but in spite of the inspector's recommendation to refuse, planning permission was granted in August 1985. Although their scheme was excluded from the original competition and summarily rejected by the Waterways Board, the Limehouse Development Group kept up their criticism. They publicized their case in *The Limehouse petition* (Wates 1986), which was signed by senior politicians and professionals as well as local residents. This conflict illustrates some of the competing views of how Docklands should be redeveloped. The Limehouse Development Group proposal was a highly imaginative mixed development scheme, based on public access to recreational facilities. On the other hand, Seifert designed a scheme which was essentially private and exclusive, for the maximum commercial return.

The Isle of Dogs

The Isle of Dogs is a peninsula formed by a curve in the Thames. It is dominated by the abandoned India and Millwall Docks which effectively cut 'The Island' off from the rest of East London. Before the LDDC took over and most of the area was declared an Enterprise Zone, it had a substantial residential population of 13 000, 90% of them local authority tenants. But by the late 1980s, and in spite of their active campaign group, the Association of Island Communities, led by 'President Ted' Johns, the future of the working-class residents was in doubt; for according to the property press, 'it is quite conceivable that in 10 years time the Isle of Dogs will have Britain's second largest concentration of office space, after the City' (*Chartered Surveyor Weekly* 4 December 1986). The transformation of the Isle of Dogs epitomizes the style of planning represented by the LDDC, which, fittingly, located its own offices on Millwall Dock.

The majority of commercial development on the Island has, not surprisingly, occurred in the Enterprise Zone, designated in 1982. Enterprise Zones provide a number of valuable benefits to firms which locate there, including ten years without paying rates, tax concessions on buildings and simplified planning controls. Small-scale development began before 1982 and produced some interesting schemes, such as Cannon Workshops, a converted and extended former dock building providing 130 workspaces; and another conversion, Limehouse Studios, an independent television production company. But by the end of 1986 major foreign and UK developers were vying for the few remaining sites in the zone, and the schemes were getting larger. There are too many individual developments to mention, but a few examples will convey the flavour. Indescon Court, 11 office suites totalling 88 000 sq. ft, was one of the first schemes to be fully let. Skylines, a so-called 'professional park', consists of 40 self-contained units aimed at designers, architects, publishers and other professional service industries. It sold immediately to owner–occupiers. Large office schemes, such as South Quay Plaza (330 000 sq. ft) and Greenwich View (300 000 sq. ft) let well, and more developers came forward. The Heron Quays development, a mix of offices and flats, was extended from its original 600 000 sq. ft to 2 million sq. ft. Other large schemes include Brunswick Wharf, Harbour Exchange and Meridian Gate. But all of these were dwarfed by the proposal, first unveiled in 1985, for 8.8 million sq. ft of offices at Canary Wharf.

The Canary Wharf project is a startling phenomenon. It represents a large increase in the total office floorspace in the capital, reflecting the extent of 'deregulation' of the private sector that has occurred in London. It is a product first of the 'Big Bang' and the resulting increase in demand for very large office suites, particularly from foreign banks and securities dealers. Office locations well beyond the City have now become acceptable, for both headquarters and backroom functions, and 'the City has become a dispersed place' (Knight, Frank & Rutley 1987, p. 10). Secondly, it is a product of the relaxation of planning controls in the Isle of Dogs Enterprise Zone. Many were astonished that a proposal of such scale and audacity could receive planning permission without even a public inquiry. It depends crucially on improved communications, and the developer offered to share the cost of an extension of the DLR into the City.

The prospects for the full development of Canary Wharf were, at the time of writing, still uncertain. However, when two members of the original development consortium pulled out, they were quickly replaced by a Canadian company, Olympia and York, and the initial Building Agreement was signed in July 1987. Described as a 'Kowloon-on-Thames' (Pawley 1986), Canary Wharf looked set to put the Isle of Dogs firmly on the global financial services map. It has been estimated that it could bring 40 000 jobs to the area, as well as doubling Tower Hamlets' rate income. Perhaps it is not surprising that the Docklands Forum, the Limehouse Development Group, and even Ted Johns, conceded some support for the development, seeing at least some chance of future employment for Island residents. As Johns put it:

> ...for the first time there might actually be jobs for local people in the enterprise zone. The Corporation has spent millions in the last four years and claims to have created 1400 jobs. We did a survey and found that only 28 were filled by local people – and half of them worked in the LDDC offices. (Pawley 1986, p. 14)

Other residents were less sanguine: one remarked, 'I think this centre (Canary Wharf) could end up being the Islanders' last stand' (Peter Wade, quoted in *The Guardian* 12 November 1985).

In addition to the financial services and wider professional services becoming established on the Isle of Dogs, the other noticeable trend is the large number of publishers coming to the area. The newspaper industry has all but moved out of Fleet Street and into Docklands. The *Daily Telegraph* print works were the first to be built on the Island, followed by those for the *Guardian* and the *Financial Times*. The *Telegraph* also moved its office headquarters, and other publishers have moved into office premises in the Enterprise Zone, including Reuters. 'With similar moves by other papers and media organisations in the pipeline, London Docklands looks set to become the media centre of the 1990s (*Chartered Surveyor Weekly* 4 December 1986).

The Isle of Dogs has also acquired its share of luxury housing developments, mainly around the southern perimeter where some flats have a stunning view of the Royal Naval College and the Queen's House at Greenwich. By far the most dramatic scheme is The Cascades, a 20-storey block of flats on Westferry Road, complete with health club, pool and gymnasium. There are also mixed developments on the Island, such as the Brunel Centre (offices, luxury hotel and restaurant) and the Chinese ICE scheme (trade centre, department store, hotel and shops). Associated Dairies opened an ASDA superstore in 1983.

The Surrey Docks

South of the river, the Surrey Docks area consists of two distinct zones of rather different character. At the western end is the North Southwark riverside, where old wharf buildings crowd the bank of the Thames. The larger eastern sector is the site of the filled Surrey Docks themselves, and the retained Greenland Dock. This area also includes some council estates which housed most of the original 9000 population. Southwark's draft local plan for the area (London Borough of Southwark 1985) was overruled by the Secretary of State in 1986, leaving the GLDP as the statutory development plan.

Upstream of Tower Bridge is London Bridge City, a large commercial development of some 2.5 million sq. ft, which includes offices, retailing, luxury flats and a private hospital. This development has a long and controversial history going back to the early 1970s (Ambrose & Colenutt 1975, Brindley & Stoker 1987). It falls slightly outside Docklands proper, and was included in the LDDC area against the wishes of its first chairman, Nigel Broackes (1984), apparently to remove the development from local planning control.

To the east of Tower Bridge are several developments which are more typical of Docklands. The Anchor Brewery has been converted into luxury apartments, many having a priceless view of the bridge itself. The Brewery site has been developed by Andrew Wadsworth's Jacobs Island Company. Wadsworth carried out one of the first warehouse conversions at nearby New Concordia Wharf. This produced 60 flats together with some office space and workshop/studio units, and was seen as a pioneering venture in the early 1980s. Between these two schemes lies Butlers Wharf, a large complex of wharves and warehouses being developed by Conran Roche. The scheme will include flats, offices, shops, studios, workshops and leisure facilities, together with the Boiler House design museum, relocated from the Victoria and Albert Museum. At the centre of the scheme is the historic street of Shad Thames which, with the other attractions, is likely to make Butlers Wharf popular with tourists. Flats in the first phase, Cinnamon Wharf, went on sale at prices between £165 000 and £325 000 in 1987 (garage spaces extra).

The major part of Surrey Docks is being developed for housing, with some industry. Some 3500 new dwellings are planned for the area, which will more than double the existing population to about 20 000. The LDDC has used its significant landholdings to produce more 'affordable' housing (less than £40 000) in this area, and it has been particularly pleased with the number of houses selling to local residents. In 1985 it claimed that 51% of 641 dwellings completed to date had sold to Southwark residents. Some blocks of council flats have been sold to private developers for conversion and sale, notably at Downtown where most of the tenants were rehoused in a new council development. To serve the increased population Tesco is developing a large retailing centre at Surrey Quays.

This area has seen some conflict over housing proposals. At Cherry Garden Pier, on the riverfront, the LDDC staged an architectural

110

competition for a private housing development. But local residents objected and eventually won half of the site for local authority rented housing. A site at Swan Road was physically occupied by tenants of an adjacent council block, with a similar outcome. As a result of these incidents and other pressures, the Corporation allocated four sites in the Surrey Docks area for up to 500 rented dwellings.

Surrey Docks was the intended site of a Trade Mart in the LDSP, but this project was abandoned. There is only one major industrial scheme, a new print works for Associated Newspapers, publishers of the *Daily Mail*. A number of existing firms have survived around the Greenland Dock but rising land values are gradually squeezing them out. This has led to the accusation that the LDDC is not doing enough to protect local jobs. In reply it claims that the rate of job loss has been slowed in Docklands and that, in any case, the area is better off without the 'backyard industries' that have moved in during the years of decline (Rotherhithe Community Planning Centre 1986, p. 26). Nevertheless, Southwark Council and the North Southwark Community Development Group have stuck by their belief in lower land values and the regeneration of industrial employment which can use the skills of the existing resident workforce: 'We are not convinced by the argument that if you didn't have good industrial buildings then people wouldn't rent them. We believe they would.' (ibid., p. 24).

The Royal Docks and Beckton

The final area of Docklands, and the last to attract commercial redevelopment, is the Royal Docks. This section of the LDDC empire comprises two distinct zones. North of the docks themselves is Beckton, a low-lying marshy area which was drained in the 1970s and has now become a residential suburb. The London Borough of Newham's *Beckton district plan* was adopted in 1980 and had some influence over the development of this area. Large numbers of family houses with gardens have been built by speculative housebuilders, mainly on LDDC land, arranged around a District Centre with an ASDA superstore. By the end of 1987 some 3700 new houses had been completed, adding considerably to Beckton's 6500 population. Initially, the LDDC found it difficult to convince the housebuilders that a market existed in this part of East London. Some were tempted with deals which linked a Beckton site with a more juicy one in Wapping, and this set the development of the area in motion. A revised *Beckton local plan* (London Borough of Newham 1986a) was put before a public inquiry in 1986 and accepted with minor modifications. Newham wanted to see more rented housing and social facilities, but otherwise supported the development of Beckton as a mainly residential area.

The Royal Docks are the most dramatic feature of the whole of Docklands. Comprising 237 acres of water, three miles in length and enclosed by ten miles of quays, the scale of these docks is breathtaking. The area also includes the residential communities of Silvertown and North Woolwich, with some 6000 existing residents. Both the LDDC and its critics have emphasized the enormous potential of the Royals for redevelopment. In the words of the Corporation's Annual Report:

> The Royal Docks is destined to become the most exciting regeneration project to be built this century. Some opportunities happen only once in a lifetime and the development potential of the Royals is one. (1986b, p. 31)

To the critics of Docklands planning under the LDDC, the Royals represent the last chance to get it right:

> All is not yet lost. To the east lies the empty vastness of the Royal Docks: a landscape to chill the soul...Demand is high now. Everyone wants a juicy slice of Docklands. What is urgently needed, in the public interest, is a plan. (Davies 1987, p. 37)

In fact, the Royal Docks have not been wanting for plans. One was prepared in 1983 by local community groups, with the assistance of the GLC Popular Planning Unit, to counter the original proposal for London City Airport. Called *The People's plan for the Royal Docks* (Newham Docklands Forum 1983), it argued for rented houses with gardens, industrial and service jobs for local people and a range of social and community facilities. It was a plan for mainly public rather than private investment, but it contained many imaginative and positive ideas for economic uses for the docks and related buildings. In 1986, Newham produced their own *South Docklands draft local plan* (London Borough of Newham 1986b), which adopted similar policies. But both plans were rejected in favour of the LDDC's Draft Development Framework of 1985, one of the Corporation's 'flexible' planning documents designed to attract private investment. The Development Framework for the Royal Docks envisaged a 10–15-year period of reconstruction, based on £150 million of basic infrastructure, mainly drainage, road improvements and repairs to the dock structures. The Corporation was seeking large integrated development schemes to match the scale of the docks. Initially, three major proposals came forward from large property development consortia, and others followed with competing plans. 'The schemes themselves are truly gigantic', enthused *Docklands* magazine, 'Nothing has been built on such a scale in the UK before' (March/April 1987, p. 27). Even the LDDC was taken by surprise and had to look for ways of accelerating its infrastructure programme. The interest reflected the unexpected early success of Wapping and the Isle of Dogs, agreement on major road links connecting the Royals with the M11 and eventually the East London River Crossing, and the construction of London City Airport on the peninsula between the Royal Albert and King George V docks. The airport is a facility unmatched by any other European city in its proximity to the commercial centre.

The proposed schemes for the Royal Docks are, at the time of writing, still being negotiated, but a few details illustrate the way private developers were thinking when faced with this unprecedented opportunity. One, by Rosehaugh Stanhope, developers of Broadgate in the City of London, consists of a gigantic regional shopping centre with up to 10 000 car parking spaces, and a 'business park' with associated exhibition and leisure facilities, amounting to some 4.5 million sq. ft in total. Another scheme, proposed by Conran Roche with Heron and Mowlem – all three active in other parts of Docklands – includes 3500 houses and flats, in various tenures. It also incorporates an information technology centre, a 500-bed hotel, offices, studios and a supermarket. A third scheme, by a consortium of British, Dutch and American developers, is a proposal for 'Londondome', a 25 000-seat stadium plus trade mart and exhibition hall, hotel and conference centre, offices and 1500 housing units. Taken together, these proposals are predominantly for multi-use business space and large-scale retailing, with a major housing component. The office content is relatively low, as the area is still overshadowed by the Canary Wharf plans, but this could increase in the future (Knight, Frank & Rutley 1987).

Following the re-election of the Thatcher government for a third term in 1987, Newham made a significant change of tactics. They decided to negotiate with the LDDC for a package of social benefits in return for their support for the Corporation's planning framework. The borough took advantage of the fact that it still owned some small pieces of land which were integral to the development, as well as other land which could contribute to it. In September 1987 it was announced that Newham and the LDDC had agreed to the inclusion of 1500 social housing units and a £10 million social development programme in the Royal Docks plan. The deal also included provision of local job-training facilities and a target of 25% local employment. The social facilities would be provided out of public and private funds, and so included a large element of planning gain. In exchange for their land holdings Newham were also seeking an equity share in the new developments which would make them a full 'partner' in the project.

Leverage as a style of planning

The LDDC has been the most prominent example of leverage planning in the 1980s, and its apparent success in its own terms has led to the wider extension of this style of planning to the regeneration of other inner-city areas. This case study allows us to draw some conclusions about the essential characteristics of leverage planning in terms of the framework set out in Chapter 2, and to comment on the particular tensions and conflicts which it engenders.

Institutional arrangements

We have described the LDDC as a quasi-governmental agency. This is a special-purpose executive agency, centrally funded and responsible to a government minister and Parliament. While there are many such agencies, the urban development corporation has become a model type of institution in its own right. The LDDC differs from most quasi-governmental organizations in being small, with a permanent staff in 1987 of only 90, plus 88 fixed-term contract staff (LDDC 1987). It does little actual planning but makes extensive use of consultants in drawing up area frameworks, as well as in many other functions such as engineering, quantity surveying, architecture, finance and marketing. Expenditure on consultancy fees totalled £10.9 million in 1986/7, more than three times in-house staff costs (ibid.). The emphasis of the Corporation's own staff is on implementation, bringing about the physical and economic regeneration of the area. Its attitudes are entrepreneurial and it has been directed by people whose roots are in large-scale property development and City finance – its first chairman was Nigel Broakes, chairman of Trafalgar House, and its second Christopher Benson, chairman of MEPC. In several of these respects the LDDC has become a model for the new urban development corporations set up in 1987.

We have seen that the main activities of the LDDC centre on the preparation of sites for private development. The key to its role as an implementing agency is land ownership. As well as the power to buy and sell land using its government grant-in-aid, the Corporation has gained control over the disposal of most of the large areas of land owned by other public bodies in Docklands.

Soon after it was set up, some 650 acres of land owned by the GLC, the Port of London Authority and the Docklands boroughs was vested in the Corporation by Parliament. There have been further vesting orders and other publicly owned sites have been acquired by agreement or compulsory purchase, making the Corporation the major landowning interest in Docklands. This has given the LDDC considerable influence over the pattern of land use in its area. By the construction of major communications routes and the provision of specific types of infrastructure on land in its ownership, and by selective land disposal, the LDDC has steered development much more effectively than it could have done using the powers of planning control alone. Many have remarked on the apparent irony of this highly interventionist role (much more interventionist than that of the Docklands Joint Committee, for example) being adopted by an agency of the Thatcher government. The *Daily Telegraph* was even moved to observe that:

> the Corporation...is more of a socialist concept than a child of a Tory administration. It is state-funded and possesses centralised planning and extensive authority to by-pass opposition. (2 February 1987)

Politics and decision-making

The main political characteristic of the LDDC is its insulation from existing local government institutions and therefore from local democratic accountability. The independence of the Corporation causes a great deal of uncertainty, and a degree of resentment, in its relationship with local government. The LDDC has one principal local government power, that of development control. It is therefore formally subject to borough planning policy, as incorporated in statutory planning documents, and housing policy. In practice, the power of development control, together with that of land ownership, has effectively overridden both the planning and housing policies of the boroughs. The Docklands Consultative Committee complained that:

> Since its frameworks are very often in serious conflict with statutory Local Plans, the LDDC is creating major confusion and resentment over the exercise of its planning functions, as it is in effect usurping the Borough's and the Greater London Council's plan making roles. (1985, p. 29)

As in so many areas of planning in the 1980s, the status of plans produced through the development plan system has been eroded and the role of local consultation, public inquiries and the entire panoply of 'democratic' planning put in doubt. Posing the question, 'What is it about the LDDC that causes so much concern?', the Rotherhithe Community Planning Centre offered this answer:

> It is the way in which it becomes impossible for any group to influence decisions of the LDDC which causes most bitterness. Their meetings are secret, their reports confidential and the results of consultations are often ignored. (1986, p. 27)

As well as being outside local democratic control, the LDDC is closely allied with the private corporate sector, in particular large property development companies. It is private developers who are carrying out most of the redevelopment of Docklands, on the basis of infrastructure provided by the LDDC. The Corporation therefore has to have a close and sympathetic relationship with developers and a detailed understanding of their investment criteria. Large projects are discussed with potential developers prior to the submission of planning applications. Although this is a common procedure in local authority planning departments, the Docklands Consultative Committee has complained that 'local authorities are excluded from important meetings with developers', contrary to the LDDC's Code of Practice. Prior to 1986, meetings of the Corporation's Planning Committee were held in private. To enable decisions to be made quickly, only 14 days are allowed for consultation on most planning applications, which makes it difficult for the boroughs to respond. According to the DCC, 'the majority of applications are therefore determined without any real accountability to the public or the planning authorities' (1985, p. 25).

The LDDC has defended itself against the accusation that it only consults with developers. It claims that the boroughs (and the GLC, before 1986) have not made use of liaison arrangements. There is some truth in this claim, particularly since Southwark pursued a policy of non-co-operation with the LDDC after 1982, although it later relented somewhat. In the view of the Rotherhithe Community Planning Centre, the result was that, 'In many important ways, there was a political vacuum in North Southwark in which the LDDC was able to operate with virtual impunity' (1986, p. 6). Tenants' associations took the initiative and formed the Southwark Tenants' Liaison Group to maintain some relationship with the Corporation. There is no doubt that the LDDC has been sensitive to criticism on this issue. It appointed a Community Liaison Officer and saw the role of its Area Teams as helping to build links with local communities. But in the end the Corporation cannot escape the fact that its consultation is more a matter of image than substance:

Of the 'undemocratic' nature of the LDDC, Mr Oliver (Deputy Chief Executive) said: 'That's clearly true. The corporation is an extraordinary arrangement for an extraordinary situation, defensible, if at all, as a temporary expedient to achieve something special in a short period of time' (*The Times* 19 April 1985)

The politics and decision-making of the LDDC can therefore be described as corporatist. Decisions are made by small élites within the Corporation itself and among large private developers and financial institutions, based principally on market criteria. The principle of leverage planning is the regeneration of a market in land and property where that market has waned or even collapsed altogether. It is this principle that produces the institutional form of the urban development corporation and which generates its corporatist style of politics. Those interests which oppose market criteria (or profitability) as the basis for planning decisions have, in effect, to be excluded from the decision-making process. Although veiled in references to the need for private-sector 'confidence', the government's case was that it was precisely the dominance of those interests – in the form of Labour-controlled local authorities – which was preventing the revival of a vigorous private market in the Docklands. The doubling of Docklands' population which is taking place, mainly by middle- and higher-income owner–occupiers, will partly change its political character and may help to convince reluctant financial institutions that long-term investment in the area is safe.

Conflicts and tensions in leverage planning

There have been many detailed arguments and local conflicts in the regeneration of London's Docklands under the LDDC. But behind them all lies one fundamental conflict which characterizes leverage as a style of planning: Who is it for? Concluding its detailed critique of the first four years of the LDDC's operations, the Docklands Consultative Committee commented that the boroughs did not share the Corporation's view of what constitutes 'regeneration': 'But the main difference centres around who is being planned for, and who is benefitting from LDDC plans and investments' (1985, p. 63). The opposing positions can be encapsulated in simple dichotomies: locals versus newcomers; and incremental change versus radical restructuring. But while these form the basis of political allegiances, the issues are not easily resolved.

The case against the LDDC advanced by the three boroughs, the GLC and many community groups, through the Docklands Consultative Committee and through individual reports, is that leverage planning does not bring direct benefits to existing residents, and frequently brings real disadvantages. They have repeatedly listed the quantifiable needs of the working-class population in Docklands, for housing – new rented housing and improvements to existing housing – industrial employment, and a wide range of social facilities which are poorly catered for. The evidence for these needs is contained in the conventional indicators which once formed the stuff of planning: numbers on housing waiting lists, numbers of disadvantaged groups, numbers unemployed, average incomes, and so on. It is not seriously contested, yet nor is it very seriously argued. For example, a report from the Docklands Forum on *Housing in Docklands*, published in February 1987, could not quote any statistical sources other than the boroughs' own Housing Investment Programme returns, which are notoriously unreliable. It is as if the case has been repeated so many times, to so little effect, that it has lapsed into a ritual incantation.

The boroughs also argue that vacant sites in Docklands could have been used to meet local needs. This was essentially the policy of the LDSP, which saw the decline of Docklands industries as an opportunity to correct the 'imbalance' between East London and other parts of the capital, for the benefit of all residents of the Docklands boroughs. In the case of housing, environment, social facilities and infrastructure this was mainly a question of public investment. The DJC's operational programme for the period 1979–83 envisaged public investment of £507 million and private investment of £109 million (GLC 1984a, p. 28). Much of the private investment (£63 m) was wanted in new industry. In spite of the continuing decline of manufacturing industry in Docklands during the 1980s, the boroughs and the community groups have stood by their belief that new industries could be attracted to the area which could make use of the existing skills of the resident workforce.

Naturally, this case has been challenged by the LDDC and its supporters. They reject the principle of needs-based planning as parochial and shortsighted. Docklands is seen as a regional and national issue as well as a local issue, bound up with the future of London – and Britain as a whole – in Europe and the global economy. Theirs is thus a wider definition of the 'community' which should benefit from the regeneration of Docklands. But they also challenge the claim that local people are being excluded. In the short term, some rented housing is being provided (and we have noted how additional sites were won in local campaigns); and a percentage of new owner-occupied houses have been selling at 'affordable' prices of under £40 000 (mid-1980s). The LDDC took over the role of the Docklands Partnership and therefore has an Urban Aid budget for social and environmental facilities. It claims that some local people

118

have been employed in the new industries; that the unemployment rate in Docklands has levelled off (male unemployment around 30% in 1985); and that longer-term employment prospects are bound to be better in expanding rather than contracting industries.

The differences of opinion over who is benefitting and in what ways come down to the two central issues of housing and jobs. The housing argument has produced a plethora of statistics on prices and incomes, and on purchases by local residents. While the Corporation has made efforts in its agreements with developers to ensure that some lower-priced housing is built on land in its ownership, prices have risen much faster than local incomes. According to the Docklands Forum, while nearly all dwellings built on LDDC land in 1981–2 sold for below £40 000, by 1985 only 43% were in this price range (1987, p. 17). Housing on privately owned sites has generally been very much more expensive. Various claims have been made about the proportion of house buyers who come from the Docklands or another part of East London and who can therefore be considered 'local'. For example, the LDDC has stated that 40% of houses built on Corporation-owned land have sold to residents of the Docklands boroughs (1987, p. 6). Unfortunately, such figures are tainted by the well known but unquantified practice of trading in addresses, whereby outsiders pay local residents to use their address and thus gain access to priority purchase schemes (Docklands Forum 1987). It is hard not to see the Corporation's claims about 'affordable' house prices as rather cynical, since at the same time it has boasted of a five-fold increase in land values across the Docklands, a factor on which its whole strategy depends.

On jobs the evidence is again inconclusive. While local unemployment may have levelled off, in the mid-1980s it remained exceptionally high. The LDDC claimed that 10 000 new jobs had been created by 1986 and that the decline in employment had been turned around. The prospect of major developments at Canary Wharf, London City Airport and the Royal Docks split the opposition, and the Corporation seemed to have won the argument by sheer force of numbers. The Docklands Consultative Committee also argued that very little effort was going into training and retraining for the local workforce (DCC 1985, p. 23), but this has been given greater emphasis with projects such as SKILLNET, a collaborative training venture involving the LDDC, the ILEA and Newham Borough (LDDC 1987).

By the end of the 1980s it looked very much as if the factual evidence of the success of the LDDC in bringing about the physical and economic regeneration of London's Docklands was overwhelming the opposition. The case of the boroughs and the community groups still tended to rest on the principles of the LDSP and a commitment to major public-sector funding. It did appear, however, that many local people would have been satisfied with just a marginal increase in the public-sector aspects of urban renewal, particularly in the areas of social rented housing and access to employment. As we have seen, in 1987 Newham decided that negotiation was a better tactic than opposition and came to an agreement on just these issues in the Royal Docks redevelopment. The LDDC, for its part, was showing some response to the opposition case by placing greater emphasis on its social programme and negotiated planning gains.

Perhaps in the end the more lasting criticism of Docklands' revival was that advanced by Colin Davies in an article entitled 'Ad hoc in the docks'. Asking, 'What kind of city emerges from such a process?', he found the answers:

Profoundly depressing to those who care about the future of European Cities. If Cities are about community, democracy, accessibility, public space, and the rich mixture of activities which creates a culture in which all can participate, then Docklands does not deserve to be called a city. (1987, p. 32)

Instead, the private-sector regeneration of Docklands is producing a collage of private realms, each barricaded behind its own security system. The social exclusiveness of the new houses and offices is mirrored by the physical exclusion imposed by the buildings themselves.

7

Public-investment planning: the Glasgow Eastern Area Renewal project

This chapter provides a case study of the Glasgow Eastern Area Renewal (GEAR) scheme. GEAR is taken as an example of 'Public Investment Planning' in a 'derelict' area with severe economic and social problems. The GEAR project was launched in May 1976 as a multi-agency partnership between a range of public sector organizations, with the Scottish Development Agency (SDA) taking on a coordinating role. The aim of the project was nothing less than to bring about 'the comprehensive social, economic and environmental regeneration of the East End (of Glasgow) and create conditions for the development of a balanced and thriving community'. The project formally ended in March 1987.

GEAR represents a particular type of planning response. It was initiated on the premise that large-scale, and coordinated public-sector action was the best, indeed the only, way to reverse the decay of the area. Launched at a time when major inner-city redevelopment projects and New Town schemes were being stopped in their tracks, by a range of financial constraints and changes in the ideological and political climate discussed in Chapter 1, GEAR represents a rare example of comprehensive public-investment planning to have survived into the 1980s. As such it constitutes a valuable case study. In addition, it will be possible to see to what extent changes in the planning climate have shifted the project's approach and how public-investment planning fares in a climate of financial constraint.

Glasgow's East End

The challenge faced by the GEAR project was a major one. The project covers an area of about 4000 acres, approximately 8% of Glasgow's administrative area. It covers the traditional East End of Glasgow, a slice of the city three miles long by two miles

121

wide, fanning out from the centre nearly to the city boundary, and includes a number of identifiable neighbourhoods such as Calton, Bridgeton, Dalmarnock, Parkhead, Shettleston and Tollcross (Figure 7.1). Since Victorian times the area has provided the base for many heavy and manufacturing industries, but by 1976 it was showing 'classic' signs of inner–city deprivation and decay. The economy was suffering and as a result there were many closures and a substantial loss of jobs for skilled and semi–skilled workers. Slum clearance and redevelopment further disrupted local communities. The population had declined to less than half of the 100 000 residents recorded in the 1951 census. Many of those left behind suffered unemployment at higher than regional averages and there was also a high proportion of elderly, handicapped and other vulnerable groups. Household incomes were low, and sickness and mortality rates high.

It would seem that it was 'the sheer scale of environmental problems' that 'distinguished GEAR from other Inner City areas'. There was 'an overwhelming impression' of 'dereliction and decay' (SDA 1979). The legacy of industrial decline and extensive clearance was large areas of derelict and vacant land. About 20% of land in the area was vacant, with 1.3 million sq. ft of disused industrial floorspace and 11.5% of all dwellings empty.

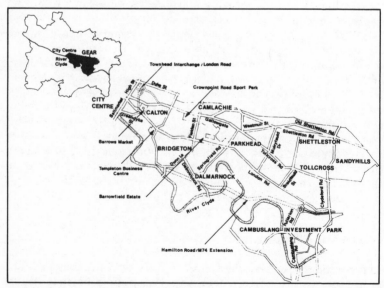

Figure 7.1 Map of Glasgow's Eastern Area, showing places referred to in the case study

The people of Glasgow's East End lived in 1976 among derelict factories, rubble-strewn spaces, abandoned railway lines and illicit rubbish tips. Their homes were in many cases in need of substantial improvement. They lived in an area that was considered to be in need of priority treatment by public-sector agencies and which had been abandoned by the private corporate sector. In short, they lived in an area of the type we have described as 'derelict'.

Origins

The origins of the GEAR project rest on a long-standing recognition by a range of public bodies of the severity of the problems faced by Glasgow. According to Leclerc & Draffan (1984, p. 336) concern in the Scottish Office (central government's principal representative in Scotland) had grown from the mid-1960s. The mid-1970s saw an increased general awareness among public policy-makers of the problems of inner-city areas. Within Scotland, Glasgow stood out if only because there was no other city with problems of the same magnitude. GEAR was a response to concern by central government in particular to be seen to be doing something about inner-city decline. The rise of the Scottish National Party (SNP), which in the October general election of 1974 succeeded in winning sufficient support to send 11 MPs to Westminster, probably also increased the motivation of the Labour government to take positive measures to help Scotland.

The opportunity to create a multi-agency response such as GEAR was provided by the reorganization of local government in Scotland in 1975. The dominant position of Glasgow Corporation was broken with the establishment of Strathclyde Regional Council, leaving the District Council with a much restricted range of functions and powers. This fundamental change in the public administration in Glasgow 'meant that the political climate ... was much more receptive to an external initiative' (Leclerc & Draffan, p. 335).

The Scottish Office began negotiations in the early months of 1976 with Strathclyde Regional Council, Glasgow District Council and the Scottish Special Housing Association (a central government funded agency established in 1937 with powers to build and manage houses). It also called in the Scottish Development Agency (SDA), a quasi-governmental agency established in 1975 with a remit to further the economic development of Scotland. The SDA was somewhat reluctant to become heavily involved, given its concern with economic as opposed to physical development, but it was seen by the Scottish Office as offering the best option in terms of coordination.

The immediate stimulus to the establishment of GEAR was the announcement in early 1976 of the abandonment of Stonehouse New Town, which made a statement of alternative action a political imperative. The abandonment of Stonehouse was itself the product of a major shift in urban policy away from a policy of decentralization by the creation of new towns to a policy of inner-city regeneration. On 2 May 1976 the Scottish Office formally announced the GEAR project. Several locations in Glasgow were considered, but the decision was eventually made for an initiative focused on the city's East End.

The SDA had not even had an opportunity to recruit staff to administer and coordinate GEAR. It was not until the autumn of 1977 that the agency had established a full multidisciplinary team – the Urban Renewal Directorate – with sole responsibility for implementing GEAR. The core of this directorate was recruited from the team created to plan and develop Stonehouse New Town. It is worth emphasizing how this recruitment pattern established a direct link within GEAR to the most prominent example of public-investment planning in the post-war period, the operation of New Towns.

In addition to the public agencies already mentioned, in the spring of 1977 the Housing Corporation and the Greater Glasgow Health Board were invited to join the GEAR project. In the summer of 1978 the Manpower Services Commission (MSC) became the eighth independent public body to join in the renewal of Glasgow's East End. All eight agencies were represented on the Governing Committee established to oversee the work of GEAR: organizational arrangements are discussed in more detail later.

The early years, 1976–9

The premature announcement of the GEAR project created some initial problems. It was difficult to create the image of a dynamic new initiative when there were no firm proposals and no detailed organizational structures. However, the project was able to draw on the well advanced plans for housing improvement and new building of the District Council and the Scottish Special Housing Association (SSHA). GEAR, as well as gaining impetus from these continuing activities, also benefitted from an early action programme of environmental improvements and small industrial workshop building by the SDA. In addition, an information and community help centre was established as a 'GEARcentre' at Bridgeton Cross in August 1977.

The second main activity of GEAR's early period was the preparation of overall proposals for the project. From the summer of 1977, for a year, ten working groups of officials from the individual agencies were established to develop policies in the following areas: population, employment, education, environment, transport, community care, leisure, shopping, housing and health. The product of these groups was eventually made public in the form of a discussion document, approved by GEAR's Governing Committee in July 1978, outlining key issues and alternative courses of action (SDA 1978). There followed a period of consultation with the community through public meetings and contacts with local groups. The community councils operating throughout the area also provided an arena for public consultation.

A detailed policy document outlining 'Overall proposals' (SDA 1979) for GEAR was then prepared by the SDA in April 1979. This was approved by the Governing Committee with some minor amendments and was accompanied by a strategy statement (SDA 1980) which identified six basic objectives for GEAR:

(a) to increase residents' competitiveness in securing employment;
(b) to arrest economic decline and realize the potential of GEAR as a major employment centre;
(c) to overcome the social disadvantages experienced by residents;
(d) to improve and maintain the environment;
(e) to stem population decline and engender a better balanced age and social structure; and
(f) to foster residents' commitment and confidence.

These objectives represent a 'filling out' of the general aim of comprehensive urban renewal established at the launch of the project. The detailed 'Overall proposals' and the strategy statement provided the basis for the remainder of GEAR's work. The ten working groups were disbanded as the focus shifted to implementation.

Implementation by the public sector

A group of public-sector agencies have been the dominant implementers of the GEAR project. With total public expenditure of £315 million (at historic prices) between 1976 and 1986, the largest contributors have been Glasgow District Council (£86 m), the Housing Corporation (£78 m), the SDA (£63 m), the SSHA (£41 m) and Strathclyde Regional Council (£32 m), with most of the remainder coming from the Greater Glasgow Health Board (£6 m) and the MSC

(£6 m) (SDA 1986). As noted in Chapter 6, a similar level of public investment has been provided by the LDDC in London's Docklands since 1981. Below we briefly describe the activity undertaken by the public-sector implementors in GEAR. We concentrate on action in the field of housing and economic development.

Housing

Over half of public spending in GEAR was devoted to housing, and considerable progress has been achieved. By 1987 roughly two-thirds of all residents were living in new or modernized homes. Three key agencies have contributed to the housing programme: Glasgow District Council, the SSHA and the Housing Corporation. Each claims to have given priority treatment to the GEAR initiative in terms of its spending, although each struggled to sustain that commitment as public expenditure constraints began to bite. The three agencies have on the whole acted independently, pursuing their own programmes within the context of the overall goals of GEAR.

For Glasgow District Council the priority has been to modernize its substantial stock within the Eastern Area. An initial programme running through to 1983 resulted in the full modernization of about one-third of its inter-war housing in the area. After 1983 this refurbishment programme was curtailed and a more limited form of improvement and repair was provided. A further one-third of the council's inter-war dwellings were treated under this programme. Thus, although much had been achieved by 1987, a substantial quantity of inter-war council stock remained to be modernized.

The reduced standard of, and shortfall in, the council's improvement programme reflects the impact of public expenditure cuts. Like many other housing authorities, from the late 1970s onwards Glasgow District Council found its ability to spend money on housing restricted by central-government-imposed constraints. After some initial building for rent, and in particular to meet special needs, the District Council was not able to fund any new building. Overall capital expenditure on housing by the council in GEAR fell from £8.7 million in 1979/80 to £3.3 million in 1984/5 (Clapham & Kintrea, forthcoming). Indeed, this rate of decline was slightly greater than that experienced elsewhere in the city. Plainly, the District Council had difficulty in preventing slippage in its programme and does not appear to have sought to protect GEAR from its financial problems.

Central government pressure to reduce expenditure impacted on the housing investment of the two quasi-governmental housing agencies operating in the Eastern Area. But they appear to have been more successful in maintaining their programmes and priorities within GEAR. In the early 1980s the SSHA was devoting about 70% of its annual capital spending in Glasgow to GEAR. In later years, however, the level of spending and the priority given to the area were reduced. Nevertheless, by 1987 the SSHA, had built over 1000 homes for rent, including a considerable amount of sheltered housing. It had also acquired, mostly from the District Council, and modernized, over 1500 houses.

The Housing Corporation has also played a crucial role in housing improvement in Glasgow's East End. For much of the period of the GEAR project it was devoting a quarter of its Glasgow budget to the area. Over 4000 tenement flats have been rehabilitated by local community housing associations funded through the Corporation. The associations were run by locally-based staff and involved local residents in their management committees. Their informal, sensitive and community-oriented operating style has justifiably attracted considerable praise.

Economic development

The SDA has played the leading role in GEAR's economic development strategy. In part, the whole programme of environmental improvement undertaken by the SDA is aimed at making the area more attractive to business. Landscaping, shopfront renewal and stone cleaning have all contributed to a dramatic change in the physical appearance of the area. Decay and dereliction was, as noted earlier, a feature of Glasgow's East End and it was, therefore, a major priority for treatment in the early improvement programmes.

Beyond this activity the SDA developed a progressively stronger interest in targetted measures aimed at retaining existing employers, allowing opportunities for growth and providing encouragement to new investment. By 1987 it had assembled over 470 acres of industrial land, including the preparation of Cambuslang Investment Park, the largest green-field site inside a city boundary in Scotland (Fig. 7.1). In addition, over 860 000 sq. ft of industrial floorspace has been provided, including 159 new advance factory units and the refurbishment of several buildings, of which the most outstanding is the Templeton Business Centre. Parts of GEAR have benefitted from Business Development Area status, and an Enterprise Trust has been established to cover the Barrows market in the inner area of GEAR. A range of grants and incentives, such as LEGUP (Local Enterprise Grants for Urban Projects), have been used to assist businesses in GEAR. The SDA's work has been supplemented by the provision of light industrial units by the Regional Council, and the Clyde Workshop Project sponsored by British Steel Corporation Industry Ltd.

127

Other economic development strategies have sought to increase residents' competitiveness in securing employment. Policies in this area were slower to take off, but programmes have gradually developed to improve levels of skill, especially among the young, and to provide support for the long-term unemployed. The MSC has played a crucial role in this field alongside the SDA. By 1987 approximately 3000 GEAR residents had benefitted from various national and local training schemes, providing a range of skills and in a few instances attempts to launch people into self-employment. Additional incentives have been provided to GEAR employers to take on trainees, such as the Training Employment Grants Scheme (TEGS) which offers a 66% wage subsidy to local companies which provide a formal training programme for newly recruited workers and can guarantee a year's employment. For the long-term unemployed, some advice services and social centres have been provided. In addition, a number of not-for-profit 'community businesses' have been launched on a small scale, of which Poldrait Community Developments Ltd is the largest, with approximately 400 local people on its books in the mid-1980s.

The impact of all this activity on Glasgow's East End has, however, been diluted by a further rapid decline in the local economy. It has been estimated that policies targetted at business interests created over 2000 additional jobs during the 1976–85 period. But some 16–17 000 jobs were lost from the area during the same period. Indeed, GEAR's rate of decline has been more rapid than that of the city as a whole (Glasgow District Council 1986, p. 24). A series of industrial closures and business contractions have undermined the impact of business development policies and further weakened the local economic base.

The other feature that clearly emerges in an assessment of GEAR's strategy is that the position of local people in securing employment has remained severely disadvantaged over the period of the project. In 1985 the official unemployment rate stood at 25%, with a male rate of 33% and a female rate of 16%. Over 2000 teenagers in the area had never worked (in a 'real' job) since leaving school. And 52% of all unemployed in GEAR had been without work for a year or longer (Glasgow District Council 1986, p. 26). In effect, the rate of unemployment in GEAR has doubled since the launch of this project.

Other areas of activity
GEAR set itself the target of 'comprehensive' renewal, and although housing and economic development consumed most of its budget other policy areas have attracted attention. In terms of the welfare commitments of GEAR the picture is one of considerable progress

towards overcoming the social disadvantages of the area, although significant gaps in provision remain. Three new health centres have been built and a broad range of social service support provided. A major campaign to increase the take-up of welfare benefits was launched in the 1980s. Owing to financial constraints on the Regional Council, education services have had to rely increasingly on Urban Aid funding. Shifts in the population structure have led to a number of local schools becoming under-utilized. It has been argued (Donnison 1986, p. 21) that insufficient has been done to provide 'second chances' for GEAR residents who left school with few or no qualifications, although the provision of a further education college in the area, scheduled to be opened in 1989, may help develop provision in this field.

Since 1976 there has been a marked improvement in the provision of community meeting places, so much so that by 1987 all GEAR residents were within half a mile of community flat, tenants' hall or other meeting place. In terms of leisure facilities, a previous large deficiency has been turned into a situation of over-provision with respect to outdoor play areas, running tracks, open spaces, football pitches and so on. Indoor provision, in contrast, has suffered from cutbacks in public programmes.

In a review of GEAR undertaken by Glasgow District Council (1986) two areas are identified where progress has been particularly disappointing. The first is in the provision of shopping facilities, where private-sector willingness to invest has been limited, no doubt a reflection of the weak state of the local economy and low income levels in the area. The second is in the field of transport. In particular, there has been a series of delays over two large-scale road schemes (Fig. 7.1) and, as a result, substantial areas of GEAR have been blighted because of the sustained uncertainty. The proposed M74 extension and Hamilton Road route are considered essential to improve the accessibility of the Eastern Area, especially Cambuslang Investment Park, to the national motorway network. The second major road project involves an extension to the inner ring road between the Townhead Interchange and Landen Road, and doubts have grown as to whether the scheme is required. Neither project is scheduled for construction until the 1990s. Again, public-expenditure constraints have contributed to this failure to make progress.

Joint initiatives by public-sector agencies

As the GEAR project matured so the public-sector participants developed a range of joint initiatives aimed at tackling the problems of the area. Thus, although most work was undertaken by individual agencies pursuing their own programmes, as relationships developed between officials from different public organizations so opportunities for joint working were seized. The range of such joint initiatives is considerable. Below we describe two of these in detail in order to give a flavour of this aspect of GEAR's working.

The first example concerns the development of a sports complex at Crownpoint Road (Fig. 7.1). Land in the area had been reserved for a school, but with the population of GEAR falling there was a realization that some additional development might be required. A member of the SDA's Urban Renewal Directorate championed the idea of a 'joint-use' sports complex, sharing facilities with the proposed new school. Negotiations with various agencies led eventually to the opening in mid-1985 of the Crownpoint Road Sports Park, offering a floodlit all-weather international athletics track and other facilities, available to schoolchildren during the day and members of the public in the evening. The project received funding from the District Council (£0.75 m), the SDA (£1 m), the Urban Programme (£0.25 m), and the Sports Council (£25 000), with the Regional Council donating the value of the land.

Another, rather different, example of joint working by public-sector agencies is provided by the Barrowfield Initiative, which was launched by the District Council in 1981. Barrowfield is an isolated housing estate of 614 houses at the centre of GEAR (Fig. 7.1). It was built between 1938 and 1947, and by the mid-1970s had a reputation for violence and vandalism. The environment of the estate helped to create a desolate image, with numerous empty boarded-up properties and a layout which combined areas of high, almost claustrophobic, density with wide abandoned street spaces. Residents in the area suffered from low incomes and an unemployment rate of over 40% in 1981. Despite modernization of the housing stock in the 1970s, the estate was one of the hardest to let in Glasgow, displaying all the characteristic associated problems for residents and local authorities.

The Barrowfield project works from former council property provided by the Glasgow District Council, refurbished with funds raised by the Regional Council. District housing officials work alongside Regional social workers providing a range of decentralized services to the area. Environmental problems are being tackled by a £0.5 million scheme funded jointly by the District Council and the SDA, although the latter's involvement is less than originally envisaged and was maintained only after political pressure from local residents and the project coordinator. Some 25 local people have been employed to undertake the environmental work. Other initiatives involving the MSC and the Regional Council's Community Resources Scheme have been used to provide opportunities on the estate. A food co-operative and financial advice service have been established to help stretch low incomes. Finally, a tenants' repair co-operative has been launched which has dramatically improved the service to tenants, as well as providing a base for the development of tenant participation in the overall management of the project.

The public sector has been dominant in the implementation of the GEAR project. The majority of investment in GEAR has involved the responsible public-sector agency getting on and implementing its schemes and proposals. There have, however, been many joint ventures in which the various participants in GEAR have combined to develop initiatives and projects.

Bringing in the private sector in the 1980s

GEAR began as a partnership of public-sector agencies. In terms of investment and activity these agencies have remained dominant but, in line with shifts in urban policy and with the encouragement of the Thatcher administration, attempts have been made within GEAR to attract private-sector investment. In particular, the SDA has sought to shift its mode of operating from land preparation and infrastructure provision to a more active business development role. As Keating & Boyle (1986, p. 111) note, 'the difficulty for the analyst is in distinguishing changes in the rhetoric by which intervention must be justified from changes in the rationale of intervention itself'. Nevertheless, from the early 1980s onwards, the SDA in GEAR and elsewhere has increasingly emphasized its work with the private sector, identifying opportunities for investment and 'wealth creation'.

The main 'success' in terms of the attraction of private investment in GEAR has been in the area of private housebuilding. Major companies such as Barratt, Bovis, Wimpey and Bellway have been attracted into GEAR during the 1980s, providing some £87 million worth of investment up to 1986, and over 1000 owner-occupied dwellings. The District Council has supported the SDA's encouragement of such schemes. Indeed, in other parts of Glasgow a range of 'build for sale' and other initiatives has been launched with the aim of widening tenure choice and providing opportunities for owner-occupation in the city.

Private builders have been attracted to GEAR sites that have been reclaimed and attractively landscaped by the public sector. Initially, land was sold cheaply by the former public-sector owners and, in some cases, subsidized with public-sector grants and underwritten by public-sector guarantees to buy any properties not sold. Another crucial factor was the Regional and District Councils' insistence on strict enforcement of green belt policies outside Glasgow's boundaries. The policy of bringing in private housebuilders has grown in strength, and a private housing market appears to have been established. During 1987 further private housebuilding was underway. This, combined with the conversion of older rented dwellings, it has been estimated, might lead to an additional 2000 owner-occupied homes.

The attraction of private-sector investment in fields other than housing has been severely limited. Moreover, in terms of industry, as noted earlier, the outward flow of investment has hugely exceeded the inward flow during the project's operation. Glasgow's 'renaissance' in terms of private-sector activity is real enough, but it is a limited and restricted affair. Total new private-sector investment in housing, industrial plant and machinery, and commercial property was estimated to be £184 million by 1986 (SDA 1986). This figure compares with £315 million spent by the public sector and includes not only completed private-sector projects, but many which are still on the drawing board.

The ending of GEAR in 1987

The final meeting of the GEAR Governing Committee took place in March 1987. It formally ended the project, but at the same time launched a 'declaration of continuing commitment' by the various participants in the initiative (SDA 1987). Further substantial public investment is promised for the future by the SDA, the Regional Council and the District Council. The other partners, too, offer further involvement and investment in the area.

New management arrangements to replace those operating in GEAR were also part of this 'continuing commitment'. Broadly, the District and Regional Councils will take over the lead responsibility from the SDA. Together they will head a Strategy Group serviced by an East End Administrative Support Unit, drawn from officials of the two local authorities. Consultative mechanisms with local residents will also be strengthened. Alongside these arrangements and focusing particularly on encouraging the local economy and business, an East End Executive will be established. This will involve representatives from the SDA, the local authorities and the private sector. Again, a locally based team of officials will support the activity of the Executive. More generally, these revised arrangements for directing the renewal of Glasgow's East End reflect the continuing dominance of the public-sector agencies.

The public-investment planning style

Institutional arrangements

We have seen that GEAR's objectives have generally been expressed in the form of broad commitments to 'regenerating Glasgow's East End'. The timescale was not made clear at the outset and it was only after a review in 1983 that the project's end date of March 1987 was formally agreed. This loose style of operation is also reflected in the institutional form adopted to run the project.

A Governing Committee was established to oversee the general direction of the project and is made up of political and leading professional representatives from the organizations involved. The committee was initially serviced by a 'consultative group' of senior officers from the various agencies, but this group was superseded in 1980 by a 'management group' of less senior officials who were more directly involved in the running of GEAR schemes. It would appear that the Committee has generally 'rubber-stamped' proposals stemming from other sources, usually the two officer groups, *ad hoc* joint working parties or individual agencies. Each of the involved organizations has retained its statutory responsibilities and powers and its formal independence. The Governing Committee has no authority to instruct participating bodies to follow GEAR policies and it has no financial resources or budget directly under its control. From a classical management perspective GEAR is a very sloppy organization. Critics point to a lack of clarity about goals, timescale and performance indicators and the absence of effective command structures to ensure coordination. Indeed, in making these points Booth *et al.* (1982) go on to claim that GEAR was and would remain a 'mango', a mutually non-effective group of organizations. They go on to describe GEAR's organizational arrangements as 'dysfunctional' and 'redundant'.

These criticisms have some merit, but they overlook the hidden, more informal bonds between the public-sector partners involved in GEAR and the particular role played by the SDA. Part of the reason behind GEAR's achievements is the shared underlying commitment of the participating agencies to do something about the problems of Glasgow's East End. The loose organizational arrangements of GEAR provided an opportunity for building on and sustaining this consensus. The vagueness of some of the project's goals may have helped, as they expressed overarching purposes and values, aimed at 'hearts' rather than 'minds'. Antagonisms emerged at times but they were never sufficient to shatter the relationship between the partners. This contrasts, for example, with experience in Liverpool during the 1980s (House of Commons 1983). A shared belief in the seriousness of the problems facing the area helped to command sustained action from all of the participants in GEAR.

The SDA also played a key role in GEAR, having a greater influence than that suggested by formal organizational structures and allowed for by Booth *et al.* (1982). It provided throughout most of the project's life a new and substantial staff and financial input. Initially, some 37 SDA staff (24 professional and 13 support) worked on the project. By 1981 the SDA had expanded its role into other cities and initiatives. As a result the structure of the agency was revised so that the GEAR team became part of the Area Development Directorate. The team was slimmed down to nine staff (seven professional and two support) because it was able to draw on the assistance of other divisions of the SDA. No other agency found it necessary to create a full-time staff resource with sole responsibility for GEAR. Instead, various officers and managers were brought in as required. These public officials from the SDA and elsewhere provided a full 'in-house' service to GEAR. In contrast to the LDDC, little use was made of outside consultants or short-term contract staff. The agency showed itself to be more adept at working with Glasgow's local authorities than the LDDC with its local councils. In GEAR, at least, it sought to complement rather than dominate its other partners. For much of the time the various participants were left to get on with their own work and responsibilities. Indeed, as noted earlier, much of the work of GEAR was carried out in this autonomous way. Thus, for instance, the community-based housing associations, which made a major contribution to the East End's renewal, found in the words of a spokesperson that 'the paths of the SDA and housing associations hardly crossed'. However, it was in preparing the overall proposals for GEAR and the development of joint ventures that the SDA's coordinating role came to the fore.

Particularly with respect to schemes at the implementation stage, SDA brought a range of very valuable skills. One agency official described it as the 'Arthur Daley approach to planning', a reference to a character in the Thames Television series, *Minder*. The main features of the approach are a willingness to broker, bend and bring together different interests and organizations, as in the example of the Crownpoint Road Sports Park:

> Somebody might have an idea. You know somebody else who has a related problem and bring the two together and smooth the path of progress. (SDA official)

Agency officials found such a role possible because responsibility for taking decisions within SDA was given to staff operating at grass-roots level. A decentralized management style meant in particular that financial support 'to smooth the path' of projects could be allocated without reference up to senior management in a great number of cases. Officials were empowered to provide funds up to £1 million. In the later years of GEAR, however, the SDA did tighten control from the top.

SDA officials operated in the context of an informal network of contacts between officials from different agencies within GEAR:

> Things happen because a whole range of officials in different agencies know and trust one another. They are more than professional colleagues and in many cases are even social acquaintances who go out for drinks together. (SDA official)

Co-operation emerges from this informal network as meetings are held and telephone contacts made. Such arrangements, together with the enthusiasm and commitment of project workers in areas such as Barrowfield, provide important factors in explaining why GEAR was at least partially effective.

In short, GEAR's organizational operation is characterized by the key role adopted by a quasi-governmental agency (the SDA) 'networking' in partnership with a range of other public-sector organizations. This organizational arrangement was not 'dysfunctional' or 'redundant'. On the contrary, it made a crucial contribution to GEAR's achievements. The very strength of these organizational relations, however, ensured that the main policy-making process within GEAR was dominated by a small number of public-sector officials, with both the private corporate sector and the local community more on the sidelines. This point is explored further in the discussion that follows of the political style of GEAR.

Politics and decision-making

As we have seen, GEAR involved various public-sector agencies in a coordinated, long-term and interventionist attack on the problems of Glasgow's East End. The rationale of GEAR's politics is provided by the perceived need to bring together the full range of public-sector implementers in order to ensure effective urban renewal. Each is invited to have a say in the running of the project in return for their co-operation and sustained commitment to the regeneration of the area. This is what gives GEAR's politics its corporatist nature. But what marks it out as a hybrid 'administrative' form of corporatism is the fact that the consultation arrangements are confined to state agencies, and do not rest on collaboration between these agencies and major organized interests in the private sector. The appropriate label

135

would appear to be state or administrative corporatism (cf. Lebas 1983, p. 9). The policy process within GEAR has been dominated in both formulation and implementation by the officials from the various participating public-sector agencies. As we have argued, for much of the time the direction of GEAR was determined by the constituent organizations pursuing their own programmes, led by their public-sector managers. Coordination was achieved through the activities of the SDA and the informal network that developed among a number of state officials.

The private corporate sector was largely uninvolved in the policy formulation process, and came in at the implementation stage in accordance with public-sector plans, and at the behest of public-sector officials. This emerges, for example, in the case of the private house-building programme in GEAR. The initiative was led by public-sector officials who established the terms and conditions of private-sector involvement.

The limited extent of accountability to the residents of Glasgow's East End bears out the corporatist label of GEAR's politics. Considerable lip service was paid at the launch of the project to the need for public participation and involvement. Indeed, one commentator (Donnison 1986, p. 21) argues that GEAR developed a 'community-based' approach. He emphasizes the extent to which politicians and officials have got alongside the groups they serve, listened to their perception of problems and given them some control over resources. Undoubtedly there were numerous public meetings, several opinion surveys, regular discussions with community councils and, since the early 1980s, two area liaison committees of officials, politicians and community leaders. On the whole, however, the public accountability of GEAR was limited, for a number of reasons.

The very size of the area covered by GEAR made a community-based approach problematic. Indeed, the 1982 social survey conducted by Professor Donnison's research team provides support for this point. When local people were asked what part of Glasgow they lived in no-one mentioned the GEAR area and only 14% said the East End. Drawing on this evidence Middleton (1985) argues that 'Its recent official designation treats Glasgow's East End as a single entity, but the area is, in fact, made up of a number of distinct communities'. The people of GEAR saw themselves as belonging to a series of distinct, traditional industrial communities. This made the concept of participating in the planning of GEAR difficult to mobilize.

A second key point is the limited commitment to participation among GEAR's public-sector implementers. The process was characterized by local people as 'information giving'. Others have argued

(Booth *et al.* 1982, p. 64) that the task of the SDA was to 'respond and react to participation demands, not to actively encourage them' while a study by Nelson (1980) concluded that participation was no more than a marginal element in the GEAR strategy and never developed beyond informing local people of plans and proposals.

Only in rare examples, such as that of Barrowfield, did the public become a partner in a GEAR project. GEAR's was a 'top-down' political style, dominated by public-sector officials concerned to achieve the physical renewal of Glasgow's East End.

Tensions and conflicts: public-investment planning in a cold climate

The experience of GEAR illustrates some of the tensions which affect the public-investment style of planning in a period of severe fiscal constraint. Such planning is always vulnerable to resource squeezes and the dictates of economic management which call for public-expenditure constraint. From the project's launch in the mid-1970s the public-sector participants in GEAR have struggled to maintain their programmes and spending levels in the area. Much investment has been achieved, but slippage has occurred in housing, leisure, road-building and other areas.

Another dilemma faced by public-investment planning is also illustrated by the GEAR case. Crucially, despite the massive public-sector effort, particularly on the industrial and economic development front, it was not possible to counteract the flows of private capital and investment out of the area. Poverty and long-term unemployment remain major problems in Glasgow's East End.

GEAR has also had to bend its priorities to meet the objectives of the Thatcher administration from 1979 onwards. Private-sector housebuilding has been brought into the area on the back of public-sector land reclamation and environmental improvement, supported in the early stages by various grants and subsidies. SDA officials in particular have been quick to 'repackage' GEAR as a form of leverage planning (Colwell 1984). Behind the new rhetoric stands the reality of the continued dominance of public-sector investment and the withdrawal of private-sector capital from the industrial field, and its limited interest in commercial and shopping facilities.

GEAR has been described 'the last of the old comprehensive renewal schemes' (Keating & Boyle 1986, p. 158). It sought to learn the lessons from previous public-investment-based clearance and redevelopment schemes. The need for coordinated inter-agency working was recognized. So, too, was the need to inform local people and support community activities. Yet GEAR has retained the fundamental 'top-down' character of public-investment planning.

The goals and direction of the project have remained firmly in the hands of public officials. GEAR has in many respects been a sustained apology by the public authorities to the people of Glasgow's East End for the havoc and destruction wrought by previous clearance and redevelopment schemes and massive deindustrialization.

GEAR has not led to a substantial change in the structure and composition of the population of the area. The new private-sector housing has attracted some 'dual-income', younger, higher-status households. The population level has stabilized at around 45 000. However, over 70% of the population still rent their housing from public authorities; 'right to buy' purchases have been minimal (below 2% of total public-sector stock); and social disadvantage, low income and high unemployment remain. The environment has been improved, leisure and community facilities provided and the standard of housing has increased substantially. Yet, at the close of the project in March 1987, one of Glasgow's leading councillors was moved to comment that GEAR's prime achievement was the provision of 'prettier street corners for the unemployed to stand on'.

8

Private-management planning: Stockbridge Village, Knowsley

Private-management planning is premised on the belief that the recovery of some of the most deprived areas in the country can be achieved by private-sector agencies taking over the management, the resourcing and the direction of the renewal process. Public-sector agencies, in particular local authorities, are seen as having failed these areas. In contrast, the dynamism and imagination released by bringing in the private sector will, it is argued, enable such areas to be brought back to life. This chapter provides a study of private-sector management in practice by examining the takeover of a problem council estate in Knowsley, Merseyside, by a consortium of private agencies including Barclays Bank, the Abbey National Building Society and Barratts, a major volume house builder (Fig. 8.1).

Knowsley and its problem housing estates

Knowsley Metropolitan Borough Council was established at the time of local government reorganization in 1974. The borough was an amalgam of smaller authorities stretching on a north–south axis along the eastern boundary of the City of Liverpool. There was no natural centre and no strong organizational base on which to build. The authority had to create, almost from scratch, a management structure capable of dealing with the full range of local authority services.

The administrative task was particularly great in the field of housing, as Knowsley inherited a large stock of 39 000 council dwellings, much of it built during the redevelopment programmes of the 1960s. The sheer size of the stock posed problems to which the authority responded with a system of area-based housing offices. Beyond this there were difficulties with the structural condition and the state of repair of much of the stock. Moreover, there were significant social problems on many estates. In 1981, compared with other districts,

139

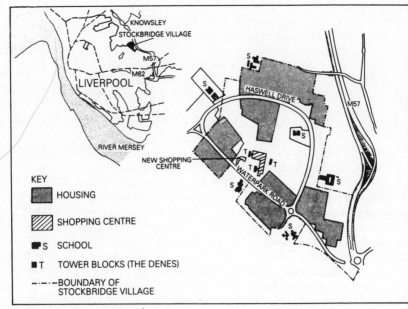

Figure 8.1 Map of Stockbridge Village, showing places referred to in the case study

Knowsley was ranked fourth on one-parent families and sixth on overcrowding.

These difficulties were compounded from the mid-1970s onwards by the impact of economic restructuring. The heavy preponderance of manual workers in Knowsley suffered massive job losses as the local firms and the branch plants of large companies closed throughout Merseyside. By 1981 Knowsley had the second highest incidence of unemployment in England. Its population was falling as those who could find employment elsewhere left. The local authority faced an increasingly difficult financial situation. Industrial decline further undermined an already weak rate base and government-imposed expenditure constraints were beginning to bite hard.

The estate at the centre of the private-sector management initiative we are examining exhibited Knowsley's problems in a particularly stark form. Cantril Farm, as it was then called, had a male unemployment rate of 49% in the early 1980s, rising to 80% among the young. People were desperate to leave the estate. It was estimated that if the rate of emigration from the estate were allowed to continue it would be predominantly vacant by the early 1990s. There was only a minimal number of 'right to buy' applications, a backlog of some 14 000 outstanding repairs, and problems of violence and vandalism.

When, on a tour of Merseyside, the then Secretary of State for the Environment, Michael Heseltine, asked to be taken to the authority's worst estate, leading councillors and officers had little hesitation in directing the Conservative Cabinet Minister to Cantril Farm, or 'Cannibal Farm' as it had been dubbed by locals.

The launch of the Stockbridge Village Trust, 1982–3

Following Heseltine's visit in June 1982 he invited Tom Baron, the Chairman of Christian Salvesen (Properties) Ltd, to prepare a scheme for the renewal of Cantril Farm. Baron had been Heseltine's housing adviser in 1979–80 and was an outspoken advocate of a wider role for the private sector in housing. Baron drew up a plan for a private takeover of Cantril Farm in consultation with Clive Thornton, the Chief Executive of Abbey National Building Society, and Sir Laurie Barratt, Chairman of Barratts.

Baron diagnosed three main causes of Cantril Farm's problems. First, its design and layout provided little privacy or defensible space. Open land around the estate was unsafe and unusable. Houses, maisonettes and tower blocks on the estate were laid out in such a way that both privacy and security were problematic. Footpaths and underpasses were little used and insecure.

Secondly, the population and tenure balance of the estate was inappropriate. Those with the ability to leave had done so, leaving behind an over-concentration of young, unemployed and problem families. This created the conditions for the high level of crime and vandalism on the estate. The absence of owner–occupiers was also considered a major problem. For Baron, owner–occupiers offered the advantage of having a 'natural' commitment to the area in order to protect their investment.

Thirdly, the estate, according to Baron, had been inadequately managed by Knowsley Borough Council. A combination of limited resources and weak tenant selection procedures had contributed to the estate's decline. The political constraints encouraged the spreading of limited resources too thinly, rather than concentrating on dealing with the problems of one area. It is worth noting that from the viewpoint of the local authority the problem was lack of effective tenant demand for the estate rather than their poor selection procedures. Moreover, given the resources the Council believed it was capable of renewing the area.

Baron's solution, however, was the takeover of the estate by a private trust which would not be fettered by bureaucracy or political accountability. The trust was to draw on private-sector resources and skills, coupled with a minimum amount of public expenditure, to renovate and revive the estate. The environment was to be made more pleasant and the image of the estate changed so that it could attract people back from Knowsley and elsewhere in Merseyside. Crucially, the tenure balance of the estate was to be shifted, with substantial levels of owner-occupation being established and higher-income groups being brought in. Barclays, Abbey National and Barratts all agreed to participate in the scheme. For Abbey National the commitment of Clive Thornton was critical. Like Baron, he believed that the private sector could and should make a stronger contribution to the tackling of such problem estates (Stoker 1985). Barclays Bank was approached by Heseltine after Abbey National had agreed to participate. Their top management were far from convinced about the scheme's viability but agreed to participate on social grounds and in order to oblige the Secretary of State. For Barratts, participation in the initiative was consistent with their increasing commercial involvement in inner-city renewal. They believed that the project might be commercially viable and, although no written commitments were given, the company's chairman, Sir Laurie Barratt, assured Baron of their willingness to participate.

The involvement of Knowsley Metropolitan Borough Council was also essential to the success of the scheme. It had to agree to sell the estate to the trust, and its involvement in the programme of renewal was required. The controlling Labour group had a reputation for being 'moderate' but was under increasing pressure locally from left-wing and Militant-influenced elements in the party, spreading out from Liverpool. To be seen to be co-operating with a Conservative government in the privatization of a council estate was a high-risk strategy for them. In the end the Labour Councillors agreed to participate because they had little faith in their authority's ability to obtain the required resources on its own for a successful renewal project. The plan proposed by Baron seemed to offer a solution to the estate's problems and rested on additional funds being provided by the Department of the Environment. Councillor Jim Lloyd, the leader of the Labour group, was particularly instrumental in persuading his party to accept the scheme. He also took a seat on the Trust's board and became part of the 'inner circle' around Tom Baron, which steered the future development of the initiative.

Baron's original plan was drawn up by the end of July 1982. In September Knowsley Borough Council accepted the scheme in principle and by November had agreed to the terms of the sale. On 5 November 1982 Michael Heseltine formally launched the scheme and at a public meeting in December local residents gave their support. They were told that if they wanted to see investment in the area then this was the only option open to them: it was an offer they could not refuse. The estate had deteriorated so badly under Knowsley Borough Council's management that any alternative appeared attractive. Moreover, the prospect of some employment opportunities stemming from the renovation and renewal was plainly of interest in an area of such high unemployment.

In February 1983 Stockbridge Village Trust was legally registered as a non-profit company. The Trust's Chairman was Tom Baron and other board members included representatives of Abbey National and Barclays, the leader of Knowsley Council and one other councillor, and two representatives of the local community. Barratts were to act as the sole contractor for all renovation and new building work undertaken by the Trust and its housing association. In addition, they were to undertake a number of build-for-sale schemes on land surrounding the estate and participate in the refurbishment of three tower blocks in the central area. The Trust's purchase of the estate for £7.42 million was financed by mortgage loans of £3 million from Abbey National, £2 million from Barclays and £2.42 million from Knowsley. In addition, Barclays provided an overdraft facility of £2 million. Knowsley Borough Council, however, was left with the outstanding debt for the building of the estate. The day-to-day management of the Trust was the responsibility of a Chief Executive and a small team of staff, some of whom were seconded by Knowsley Borough Council.

The task of the Trust was to take over the ownership, management and renewal of the Cantril Farm estate. This consisted of 3109 dwellings made up of 1227 two-storey houses, 1056 flats in nine high-rise blocks and 826 two-to-four storey flats and maisonettes. In addition, there was a run-down shopping area with an underground car park, a library and other limited community facilities. The aim was to remodel the estate over a five-year period, with renewal work being completed by March 1988. Baron's dream was to transform the estate into 'a place where people will want to live from choice, not economic circumstances'. According to Heseltine, the Stockbridge Village Initiative was 'a potentially trail blazing venture... nothing like it has ever been tried before' (both quoted in Grosskurth 1984, p. 25).

The implementation of the first programme phase, 1983–5

Baron's approach to the renewal of Cantril Farm was premised on a rapid boost to confidence in the area. This would create commitment among existing residents and attract higher-income newcomers. The first step in changing the image of the estate was its rechristening as Stockbridge Village. The fact that it was 'under new management' by a private trust was also actively promoted. The confidence-boosting strategy, however, rested fundamentally on an ambitious first phase of physical redevelopment to be completed by mid-1985. In addition, a restructuring of the management and repairs service provided to tenants was a top priority in order to make a substantial improvement in the quality of service provided. Taken together, these were to contribute to a shift in the social composition and community spirit of the estate which Baron considered essential to the success of the initiative. By the end of the project it was hoped that over 50% of the estate's population would be owner–occupiers.

Phase I of the development programme (April 1983 – April 1985) contained an ambitious range of projects. Among the first to be completed by the Trust was a new shopping parade of 14 small units and a supermarket (Fig. 8.1). This replaced the vandalized concrete warren which had previously acted as the area's commercial centre. Various environmental works were also undertaken, including the completion of a perimeter footpath and major landscaping improvements. Tree and shrub planting throughout the estate began to make an impact, although the rate of planting had to be stepped up to compensate for the number of plants which 'disappeared' overnight.

A further area in which substantial progress was achieved involved the building by Stockbridge Village Housing Association (under the control of the Trust) of 141 new houses and bungalows on vacant land on the estate, which were made available for rent or shared ownership. Barratts also completed 56 new houses for sale on the estate.

Early success was also achieved in the remodelling of some of the housing. This involved the demolition of some maisonettes, while some houses were 'turned around' and front door porches built onto what were previously their backs. This created greater privacy, and also greater security. In line with architectural trends the theme of creating 'defensible space' was pursued, with housing being broken up to form small neighbourly 'clusters'. Low-level walls were built to separate the small groups of houses, creating areas of privacy but allowing people to see what was going on. Tenants and residents who benefitted from these initial schemes were generally pleased. And one correspondent at least was quick to herald a 'transformation' of the estate, commenting that remodelling had given it a 'surprisingly villagey' look (Morton 1984).

The Trust also achieved substantial progress in basic repair and maintenance work. Between April 1983 and April 1985 deferred maintenance work on over 1000 properties was completed. Generally, the standard of housing service provided to tenants was improved. Again the Trust adopted an approach which was already being experimented with on other problem estates (Power 1987). A smaller-scale and more streamlined housing management service was established. A system of four area-based, decentralized estate managers was introduced, alongside a group of nearly 20 caretakers to supervise the high-rise blocks. Each manager had responsibility for just under 600 properties. The estate managers, all previously local authority staff, were given considerable discretion. Each had a budget for repairs for his/her area, could influence where the money was spent, make orders for work to be carried out and authorize payments for completed work. The repair work itself was carried out by a private contractor, who also provided an emergency service for the estate.

Each manager was set targets for reducing the number of 'voids' or empty properties and collecting rent arrears, and improvements were achieved in these two areas. Moreover, tenants found their new landlord highly satisfactory. The number of complaints about housing repairs at the local Citizens Advice Bureau dropped off dramatically. A survey conducted in 1985 revealed that nearly three-quarters of all tenants believed that the Trust provided a better housing management service than Knowsley Borough Council had previously offered. There was also evidence of a growing desire on the part of tenants to stay on the estate, and an increase in the number of applicants on the waiting list for the estate.

There were, then, signs of substantial progress during the first phase programme. The Trust had generally created the impression that a private agency could achieve more, had more money and was more efficient than the council. The Trust had fostered a lot of goodwill among residents and the redevelopment of the area had begun. But although confidence had been boosted there was no great influx of higher-income newcomers. In fact, the population structure of the estate was to a large extent unchanged, with only a small increase in the number of owner-occupiers (from 6 to 13% of the total households). The overall income level on the estate remained low. Moreover, behind the scenes the Trust had run into a number of a major obstacles which came to the fore in mid-1985.

Crisis, 1985

Stockbridge Village Trust 'hit the rocks' in mid-1985. Serious delays in the redevelopment programme became increasingly evident and problematic. In September 1985 the Trust wrote to the Department of the Environment stating that the future financial viability of the project was in doubt. Local councillors, party activists and one of its own directors criticized the Trust, claiming that it could not solve the estate's problems, and consultants were called in to investigate the Trust's position. In this section we examine the collapse of the Trust's original plan and the reasons behind its financial failure.

Delay and disaster in the renewal programme

Along with the successes of the redevelopment programme there were a number of areas where progress was slower than hoped for or, worse still, effectively blocked. The 'remodelling' of the housing on the estate was considerably behind schedule by April 1985. The maisonette demolition programme was at half the level that had been planned. Worse still, the modernization and refurbishment programme promised to tenants was running substantially behind. By April 1985 only 14% of the dwellings due for treatment had been dealt with.

Barratts' programme of new build–for–sale on land surrounding the estate was also way off target. The first development of 56 houses had been on a vacant site, detached from the remainder of the estate, and prominently situated at the entrance to the area. In short, it appeared to be the most promising of the available sites, yet the pace of sales was relatively slack. Although all but two of the properties were sold, Barratts plainly felt that the experience indicated that the market for private housing in Stockbridge Village was very weak. As a result, although they started work on a further 42 properties, they were reluctant to undertake the level of new build envisaged in the Trust's plans. The 56 houses completed by April 1985 contrasted with the target figure of 600 new houses to be built for sale by Barratts.

Barratts' involvement was also crucial to the renovation of the central area of Stockbridge Village. This was dominated by three 22-storey tower blocks, known as The Denes (Fig. 8.1). These blocks were to be decanted of their existing tenants and then sold to Barratts for £1 million. Barratts were to refurbish the flats for sale to young, mobile professionals. The underground car park beneath The Denes was to be given over to their exclusive use and they were to be serviced by a range of shops and a small leisure/keep-fit complex in the central area.

Progress was made on decanting the existing tenants from the blocks by the Trust and Knowsley Borough Council, but Barratts became increasingly uncertain about the viability of the scheme.

When it became clear that even with the aid of an Urban Develop-ment Grant the scheme could not be made to 'stack up', early in 1985 Barratts withdrew. The failure to proceed with the Denes project effectively blighted the remainder of the plans for the central area. The shops, the leisure/keep-fit centre, the health centre and other ancillary development could not proceed.

The collapse of these central area plans was a lethal blow to the Trust. Barratts' withdrawal meant not only the loss of the £1 million of sale revenue but also the prospect of considerable cost in dealing with the high-rise blocks in some way. Moreover, the refurbishment of The Denes was at the heart of the confidence-building renewal plan. The Denes project was to bring in higher-income groups, introduce a further 510 owner–occupiers to the area and provide an appropriate catchment population for the new shopping centre. By mid-1985 it was clear, in the words of Knowsley's Chief Executive, that 'there was going to be no yuppie invasion in Stockbridge'. A central element in Baron's renewal strategy was not going to come to fruition.

Financial collapse

By the autumn of 1985 the Trust also had to admit that its financial strategy was in ruins on both the revenue and capital side. On receiving news of this, in September 1985, the Department of the Environment called in consultants to investigate the Trust's difficulties (Knowsley Metropolitan Borough Council 1986). They confirmed that the position was grim and outlined a number of the reasons behind the Trust's financial failure.

On the revenue side the existing rent levels on the estate were too low to cover management and other costs. The likelihood in the future of a rise in maintenance costs as the renewal programme was completed only made the picture look more bleak. The plan had been for a high standard of maintenance and repair, to be paid for by higher rents and more efficient management. But this had completely disregarded the impact of fair rent legislation.

On properties it had improved, the Trust asked for large rent increases which would allow it to recover a considerable part of its capital investment. Generally, however, only small increases were allowed as the Rent Officer made his/her judgments on different criteria, namely comparable local rents for a similar type and quality of dwelling. Moreover, the Trust had inherited relatively high rent levels from Knowsley Borough Council and a few tenants in unimproved houses had in fact been able to achieve reduced rents on application to the Rent Officer. If all tenants had insisted on registering their rents the Trust's financial position would have collapsed even sooner. As it was, it is clear that fair rent legislation would not allow the Trust to charge the higher rents it required to support its housing services and maintenance programme. Given the nature of fair rent law there was a fundamental flaw in the Trust's financial accounting on the revenue side.

Parallel problems beset the capital side of the programme and it became clear that the Trust's funds were inadequate to complete the improvement of the area. It had been envisaged that the capital programme would be financed via property and land sales, with any shortfall being made up by public funding and a bridging overdraft. The Trust's problem was that while the public sector contribution was forthcoming the funds to be generated from the private sector were not. In addition, the costs of various projects had spiralled. Essentially the Trust had massively underestimated the cost of work and substantially over-estimated its receipts from sales by a similar amount. By late 1985 the result was an imminent financial collapse, with the £2 million overdraft facility provided by Barclays about to run out.

The consultants found that unit costs for the improvement of low-rise dwellings on the estate were by 1985 twice as high as had been originally estimated. In the case of high-rise dwellings (excluding The Denes) unit costs for improvement had spiralled six-fold. The new shopping centre was constructed at a cost of £1.1 million, compared to an original estimate of half that amount. The consultants suggested that these spiralling costs reflected a 'woefully inadequate' initial survey of the estate. The condition of its housing stock was only 'cursorily assessed'. This in turn reflected the haste with which the Trust had been launched as a political initiative.

The failure on the capital receipts front flowed from the limitations of the 'right to buy' drive among the existing residents of the estate and the unwillingness of Barratts, described earlier, to undertake their build-for-sale commitments. With well over 70% of tenants receiving full or nearly full housing benefit, the consultants concluded that the Trust's plan for large numbers of existing residents to purchase their own houses was a non-starter. After an initial surge the rate of 'right to buy' sales had dropped, and by 1985 a total of just 400 had been achieved compared to a target of 600. The level of future sales would continue to be very low, the consultants predicted, because of the lack of purchasing power on the estate.

A number of other factors were identified by the consultants as contributing to the Trust's financial difficulties. These included the collapse of The Denes project which, with the withdrawal of Barratts, turned the three central tower blocks from a £1 million asset to a £2 million liability overnight, given the estimated cost of their renovation or demolition. The sudden imposition of VAT on building works also caused difficulties, although the Department of the Environment adjusted its grant contribution to cover this Treasury-imposed cost. More generally, the high rate of interest placed strains on the Trust's financial position. One final criticism was the somewhat loose accounting control exercised by the Trust over Barratts as the sole contractors for the renovation and building work on the estate. The consultants were disconcerted to discover the informal nature of interactions and that no final accounts had been agreed for work completed. Indeed, they recommended that in future the Trust's contract work should be put out to tender and that more formal and sharper accounting procedures should be established.

The process of salvage, 1986 onwards

The private and public partners involved in the Trust reacted differently to the financial crisis. The former eased themselves out or sought to minimize their role, while the latter became more deeply committed in directing and resourcing the Trust's programme. In the discussion below we examine the position of the different participants. Attention is then focused on the revised proposals eventually agreed in September 1986 after a year of wrangling. Finally, a brief assessment is made of the progress on the reformulated scheme.

The reaction of the partners

Both the Abbey National Building Society and Barclays Bank made it plain that they were unwilling to provide any further financial backing to the Trust. They were prepared to delay repayment of their original funding but could see no prospect of additional finances coming from themselves or any other private-sector source.

Barratts, as we have suggested, were very unsure about the potential future market for private housing development in the area. It was clear that they would not complete their original commitments and the Trust had no lever with which to influence them. The sole-contractor status enjoyed by Barratts for work on Stockbridge Village was supposed to underwrite their risks in build-for-sale schemes. Yet this was an informal arrangement and there was no legal agreement or contract which the Trust could use to bargain with Barratts.

The unwillingness of the private sector to come to the rescue left the onus on the key public-sector participants. Would they let the Trust be terminated? At first sight it would appear that for the Conservative government such a prospect would be an anathema, since Stockbridge Village Trust was one of their flagships. Apparently, however, because it was a Heseltine initiative there were some in the Cabinet who would not have been unhappy to see Stockbridge Village fail. But the overwhelming pressure was for a rescue to be mounted. Such a high-profile privatization initiative could not be allowed to collapse.

The leadership of Knowsley Borough Council, too, had little desire to see the Trust publicly fail. The Council had given its backing to the scheme and its reputation depended on the project succeeding. However, a number of councillors, party activists and officials, felt that the Trust's problems provided a useful weapon with which to attack the Government's housing policy and demonstrate the inadequacies of private-sector solutions. Yet having given its support thus far it would have been difficult for Knowsley to withdraw. Somewhat reluctantly, the Borough Council found itself committed to propping up the ailing Trust.

The rescue package

After nearly a year of negotiations and discussion between the partners, a revised financial package and reformulated renewal programme was agreed in the autumn of 1986. The financial package consisted of total additional funding of about £10 million from the Urban Programme, with central government and Knowsley splitting their support on a 75 : 25 basis. Broken down, this funding included £3 million to complete the refurbishment of the estate's housing and environment, £2–3 million to demolish The Denes and £3–4 million for the building of a new swimming pool and leisure centre. The Trust's lenders – Abbey National, Barclays and Knowsley – also all agreed to defer repayments on their mortgage loans and, with effect from 1 June 1986, accept a reduction in their interest payments.

The revised renewal plans had at their core a new set of proposals for the central area of the village, prepared by Knowsley Borough Council. This involved the demolition of The Denes and the building of a new 'Caribbean-style' leisure centre and swimming pool which, it was hoped, would prove to be a major attraction, bringing people into Stockbridge to shop and as a consequence support local business. Knowsley Borough Council took over responsibility for both the demolition scheme and the leisure centre. The revised central area plan also allocated space for a new health centre (to be funded by the health authority at a cost of £0.75 million), some further shopping units and community facilities.

The renovation programme for the estate's housing remained a mixture of demolition of the worst of the maisonettes and refurbishment of other housing. But the speed of the programme was slowed and the standard of renovation work was reduced compared to that applied in the original schemes. Work was planned to be completed by 1990. Finally, Stockbridge Village Housing Association was to complete a programme of new build-for-rent and Barratts too, it was hoped, might be coaxed into undertaking further build-for-sale.

In effect, the rescue package confirmed the Trust's abandonment of a private-sector-led recovery for Stockbridge. Baron's original plans had envisaged 15% public-sector financial support, with the remaining resources coming from the private sector. By the time that the Trust had been formally established in April 1983, the ratio of planned funding had moved closer to an equal split between the two sectors. The financial package for the modified Stockbridge programme involved the public sector taking the dominant role, providing at least £2 for every £1 of private sector finance. This dependence on public-sector funding is only likely to increase in the future, as Abbey National and Barclays attempt to reduce still further, or to end, their involvement in the project.

Progress on the revised programme

By mid-1987 the Trust was able to report steady progress on its housing renewal programme. Over half the remodelling programme had been completed; approximately 1300 units had been refurbished and 658 maisonettes had been demolished. Moreover, the Trust had succeeded in obtaining substantial rent increases for some of its improved property. Stockbridge Village Housing Association had virtually completed its programme of 277 new-build dwellings. Barratts, however, had made little further progress. Indeed, about half of the owner-occupied homes that had been built were in the process of being repossessed, with the building societies/estate agents finding it difficult to attract new purchasers.

The central area redevelopment had unfortunately run into major difficulties. One setback was a fire in February 1987 which extensively damaged the shopping centre built by the Trust shortly after its launch. Work started on the demolition of The Denes in October 1986, but in mid-1987 a major industrial dispute, which at the time of writing had not been settled, stopped work on the site. This effectively blocked progress on other elements of the central area plan. There were also doubts about the viability of the leisure centre proposal. The revenue costs in maintaining and running the centre would fall on Knowsley Council in a period when it is increasingly going to be hard-pressed on its current spending budget.

Private management as a planning style

Stockbridge Village is one of the most long-running and large-scale of recent private-management planning initiatives. Our case study of Stockbridge enables us to examine in depth the institutional arrangements, the mode of decision-making and politics, and the dilemmas and tensions associated with this planning style.

Institutional arrangements

The key institutional device used in the Stockbridge scheme is the Trust. This was supplemented by a registered housing association controlled by the Trust through overlapping directorships. The Trust was seen as necessary in order to create an agency free of political control, able to access both public and private funds, and also to boast that it was bringing Stockbridge 'under new management'.

Stockbridge Village Trust is a 'non-profit-making distributing body'. It is a company limited by guarantee, and not having a share capital, whose powers, functions and responsibilities are governed by a Memorandum and Articles of Association prepared under the Companies Act 1948–81. The objectives laid down for the company are broadly the renewal of the Stockbridge area. The Trust has a wide range of powers to hold and develop land, borrow and invest funds, employ staff, and dispose of property. The liability of each member of the company is limited to £1.

The original membership of the Board is set out in Table 8.1. In a few cases the particular personnel involved have changed, but the general composition of the board has not, with the exception

Table 8.1 The board membership of Stockbridge Village Trust, 1983

T. Baron		Chairman
Cllr J. Lloyd	(Knowsley MBC)	Deputy Chairman
Cllr M. Foulkes	(Knowsley MBC)	
C. Thornton	(Abbey National Building Society)	
T. Smith	(Barclays Bank)	
J. Lawler	(Knowsley Parish Council)	
J. Everett	(Elected by tenants, April 1983)	

Source: Stockbridge Village Trust, Annual Report and Accounts 1983–4.

of the retiring Chief Executive of the Trust who was asked to join the board in February 1986. Mrs Joyce Everett was elected on a once-and-for-all basis as a community representative by way of an estate-wide poll held in April 1983. She worked as an advisor at the local Citizens Advice Bureau. The other community representative was appointed on the basis of a nomination from Knowsley Parish Council.

The Board is not formally accountable to any other party, and members' duties are to promote the activities and interests of the Trust, not to act as delegates for their parent body or constituencies. Like other companies, the Board's discussions are held in private and its minutes are confidential. Meetings are held every two months. Most of the crucial decisions appear to be made in a financial subcommittee of the Board, comprising the Chairman and the representatives of the mortgagees (Abbey National, Barclays and Knowsley Borough Council). In effect, this excludes the community representatives and it appears that many proposals and plans are first agreed by the finance subcommittee, with the full Board formally ratifying them.

The day-to-day management of the Trust is the responsibility of a Chief Executive. As noted earlier there is a 30-strong housing management staff. In addition, the Trust has had the services of a planner, an architect and legal and financial advisors and a small clerical team. Stockbridge Village Housing Association provides the Trust with a further instrument. It is controlled by the Trust through overlapping directors and the sharing of staff.

Although the Trust is formally an independent private body it does have very close working relations with a number of public agencies. Knowsley Borough Council has provided not only financial aid but also administrative support by seconding staff and undertaking other tasks such as the preparation of the central area plan, the processing of Urban Programme grant applications and the sorting out of planning and building consents. Under the terms of the conveyance when Knowsley sold, the Council also agreed to rehouse tenants displaced by the Trust's development programme, to provide mortgages to tenants exercising the 'right to buy', and several other measures. There can be little doubt that without the active support and involvement of the local authority the Trust would not have been able to operate.

Other public-sector backers have been involved on a less detailed and regular basis but nevertheless were central to the operation of the Trust. Officials from the Department of the Environment, in particular the Merseyside Taskforce, participated in the negotiations to launch the Trust and, especially after the financial crisis in 1985, have kept a watching and guiding brief over the project. The Housing Corporation's commitment has also been essential. The work of Stockbridge Village Housing Association was for the Corporation its largest programme in the country on a single estate.

Stockbridge Village Trust, then, constitutes an ambiguous institutional device. It is a formally independent agency dominated by private-sector representatives charged with achieving the renewal of a housing estate in Knowsley. But much of its funding, administrative back-up and policy advice comes from the public sector. This in turn reflects its rationale which is to achieve the private management of public policy.

Politics and decision-making

The decision-making style of Stockbridge divides into two forms. Among representatives of the private-sector participants and the public-sector funders an informal and mutually supportive relationship was established. We have already noted that the consultants brought in after the 1985 crisis felt that this informality contributed to the slackness and inadequacy of the checks exercised over the work of Barratts. On the other hand, the shared perceptions and even friendship among the Board members and other key participants helped not only to launch the scheme but also to keep it going after it had run into difficulties.

In its relationship with local residents the Trust's decision-making style took a different form. It provided information, it sought to bring people along, but it did not regard itself as accountable to local people, nor did it actively seek their involvement in decision-making. The tone of this relationship was set on the launch of the project. At public meetings and in newsletters residents were offered the opportunity to have their area taken over by a private trust, but it was made plain that no alternative was in the offing – the choice was something or nothing.

In its dealings thereafter with residents, the Trust's approach is outlined in its first annual report where it commits itself to consulting 'with our tenants to ascertain that what we plan to do has their general support and approval'. Indeed, regular newsletters and street meetings have sought to keep people informed, but there is no commitment to participation. The community representatives are there to add legitimacy but, as noted earlier, not to act as delegates from their constituencies; rather, they are meant to represent the Trust's views to residents.

A survey of residents conducted by the Department of the Environment's consultants in 1985 revealed that few tenants saw themselves as involved directly with the Trust and most claimed a very limited understanding it. Some 85% could not name their community representatives and 75% did not know who was on the Trust's board. Only 22% claimed to have been involved in consultation about plans to improve their block or area. Finally, over two-thirds were unsure about the range and scope of the Trust's activities and responsibilities.

The non-involvement of local residents in the early years of the Trust was, indeed, an explicit policy approach. Tom Baron argued that, in order to get the redevelopment programme going, strong specialist leadership from professionals and private-sector managers was required; the community were to take a back seat, becoming more involved once the renewal programme was completed. At this stage the balance of the population would have shifted, with over 50% owner–occupancy, providing an appropriate base for more participation of local people.

The Trust's relationship with local residents can best be described as 'paternalistic'. Private-management planning is about helping those whom its advocates believe are not capable of helping themselves. People are consulted more to gain their assent or acquiescence rather than to ascertain their views. What is required by this planning style is that local people are brought along, but their participation in the process is not seen in the positive way it is in, for example, the case of popular planning.

Conflicts and tensions in Stockbridge Village

Stockbridge Village Trust was born out of a political response to inner-city social unrest and riots in 1981. It was carried forward on a wave of rhetoric. Heseltine wanted to see what a 'privatization' initiative could do. During his tour of the estate he commented: 'It would be interesting to see what can be done here without recourse to public funds' (quoted in Morton 1984). The private sector was to be invited to 'save' the estate, by transforming its physical fabric and social make-up. Stockbridge Village therefore expresses a confidence in the ability of the private sector to out-perform the public sector. Private-sector management skills, vision and resources were to succeed where public authorities had failed. The speed with which the project was launched and its ambitious aims reflected the belief of its government backers in the inherent superiority of the private sector.

However, behind this rhetoric and ideology there was considerable confusion and ambiguity, which in turn contributed to the initiative's troubled history. The private-sector participants invited in by Heseltine and Baron agreed to participate not on commercial grounds (with the possible exception of Barratts) but for social/political reasons and as a favour to the government. They saw public money as not only essential to the scheme but regarded central government as its ultimate guarantor, limiting their liability and responsibility. Crucially, they did not apply commercial criteria to their investment or an assessment of the scheme's potential. They viewed Stockbridge more as a public relations exercise than serious housing policy. Government ministers, Baron and Department of the Environment officials, trapped by the scheme's ideology and rhetoric, saw Stockbridge Village as a private initiative and believed that private-sector controls and imagination would ensure the efficiency and success of the project. Despite its substantial financial support for the project, the Department of the Environment made little effort to monitor or oversee it. In short, the government and the major private-sector backers regarded Stockbridge Village as each other's baby. This, according to the consultants brought in after the 1985 crisis, was a crucial factor in explaining the project's failure, since it contributed to the grossly inadequate initial evaluation of the estate and the redevelopment plans, and to slackness in the financial management of the project.

What the Stockbridge Village initiative illustrates is that in an area of high deprivation and severe physical and economic problems it is unrealistic to expect the private sector to act as sole agents of renewal. The scheme was fundamentally constrained by the low income levels of the existing estate residents and the weakness of the local Merseyside economy. There were no mobile, higher-income groups to be drawn into the area. Heseltine's comment about 'no recourse' to public funding was always optimistic, but the history of Stockbridge Village has made it appear absurd. As we have seen, Baron's original estimate of only 15% public funding to 85% private funding has in practice come close to being reversed. In terms of our typology, Stockbridge Village has moved from the status of private-management to public-investment planning.

In more recent proposals in the mould of private-management planning, the government appears to have drawn on some of the hard lessons from Stockbridge Village. Such schemes might work in less hard-pressed areas. The government has encouraged the formation of another trust in Thamesmead, a large former GLC estate. Here there seems considerable potential for private-sector housing development on vacant land. 'Right to buy' purchases are likely to be at a high level. Moreover, there is the option of cross-subsidization of rents, with a number of viable commercial and industrial premises included in the Trust's portfolio.

In areas where the market is more depressed, however, there is plainly a need for major public support for any investment strategy and a recognition that rents have to be freed to facilitate the long-term viability of such schemes on the revenue side. The Housing Action Trusts (HATs) proposed in the 1987 election manifesto are directed at areas with major problems, but are premised on the idea that the private-sector agencies drawn in to take over these areas will have access to substantial public-sector funds for their capital expenditure programme. Once redevelopment has taken place, the HAT will sell most properties for owner-occupation. For those remaining, a landlord relationship will be established, but on the basis of 'assured' tenancies. This will give freedom from existing rent controls and enable higher rents to be charged. Of course, the strategy also implies what was not achieved in Stockbridge Village, namely a substantial shift in the nature and composition of the population of these run-down estates. In the light of this, future trusts may be established in areas where the surrounding local economy is sufficiently buoyant to provide a pool of higher-income groups who can be drawn into the area. This argument is born out by the Government's announcement in July 1988 that three of the six initial HATS will be in London.

9

Six styles of planning in practice

Reading the six case studies, one is inevitably struck by the widely differing processes and outcomes which characterize planning in the Thatcher years. While not attempting to provide a comprehensive account of planning over this period, the case studies cover a wider spectrum from the highest to the lowest degrees of state intervention and control, and from the smallest local communities to large corporate institutions. There are marked contrasts between the styles in terms of how they are operated, their effectiveness and their outcomes. This implies that each style faces rather different problems of legitimation and therefore appeals to different political constituencies and ideologies.

The purpose of this chapter is to compare the general features of the six styles of planning discussed in the case studies. While the detailed studies give depth to the picture of planning in the 1980s, the comparative analysis provides some breadth and helps to put the different styles in perspective. The analysis draws on the three broad categories used in discussing each individual style, namely institutional arrangements, politics and decision-making, and conflicts and tensions. A simplified summary of the analysis is provided in Table 9.1.

Institutional arrangements

Each style of planning has its characteristic institutional arrangements. Three general types of institution responsible for the formulation and implementation of land-use plans and development are represented in the case studies: local authorities, quasi-governmental agencies and neighbourhood-based organizations. Local authorities feature as the principal agents in regulative and trend planning.

Table 9.1 Characteristics of the six planning styles.

Planning Style	Institutional arrangements	Politics and decision-making	Conflicts and tensions	
			Limiting factors	Principal interests to benefit
Regulative Planning	Local authorities	Technical–political	Strength of local market	Local land and property owners
Trend Planning	Local authorities	Non-strategic gatekeeping	Retaining any control of market	Incoming developers
Popular Planning	Community organizations	Imperfect pluralist	Community control of resources	Local lower-income groups
Leverage Planning	Quasi-governmental agency	Corporatist	Potential for market revival	Incoming developers
Public-Investment Planning	Quasi-governmental agency	Administrative corporatist	Long-term resource commitment	Local lower-income groups
Private-Management Planning	Private trust	Paternalistic	Inability to move beyond tokenism	None

Quasi-governmental agencies have been created for leverage planning and public-investment planning. Popular planning and private-management planning are both organized by neighbourhood-based agencies, a range of community organizations and a private trust respectively. In the discussion which follows we begin by examining some of the distinctive characteristics of the institutional arrangements found in our six case studies, and go on to discuss the trend away from the local authority domination of land-use planning.

Regulative and trend planning operate through elected, multifunctional local authorities. Nearly all local authorities have a separate planning department or section, staffed by professional planning officers. The planning function is overseen by councillors through a planning committee and associated subcommittees. The Chief Planning Officer may have substantial delegated powers to take decisions on relatively minor planning applications, but major planning decisions and strategic issues require the approval of councillors through the committee system. Planning responsibilities are shared between the upper and lower tiers of the local authority system. County councils (in Scotland, regional councils) are responsible for structure plans, while district councils are responsible for local plans and the administration of development control. (In London and the metropolitan areas, following the abolition of the GLC and the six metropolitan counties, the upper tier now consists of a joint committee of councillors from the lower tier districts or London boroughs.)

The division of land-use planning responsibilities has been one of the prime sources of tension between upper- and lower-tier authorities (Leach & Moore 1979, Alexander 1982). In the case studies of Colchester and Cambridge we noted a number of conflicts between county and district authorities, as well as conflicts between neighbouring district authorities. These conflicts are often resolved at central government level, through planning appeals or calling-in procedures, but with little or no systematic regional or national coordination. In the absence of such coordination, local authorities in the South East region formed a joint committee, the London and South East Regional Planning Conference. This committee has had some limited success in agreeing overall strategic planning objectives, and the Thatcher government is considering whether to encourage other local authorities to follow this example.

Public-investment and leverage planning are both carried out by quasi-governmental organizations, but these bodies have little in common as institutional forms. The SDA was created in 1975 as a permanent organization directly funded by central government through the Scottish Office, with some provision for self-financing through land sales. Its work is overseen formally by a management board appointed by the Secretary for Scotland, and on a more informal and frequent basis by civil servants from the Scottish Office. It has a staff of over 700 and its responsibilities for economic and environmental development extend over most of Scotland. As in the case of the GEAR project, most of the work on these initiatives is undertaken in-house rather than by outside consultants. There has, however, been more use of consultants in recent area initiatives (Keating & Boyle 1986).

The SDA operates through a system of functional directorates and area teams. It constitutes a major bureaucracy with concerns which range from property management through small business creation and urban renewal, to economic development initiatives for particular industrial sectors. The agency has a complex management task in monitoring and coordinating its various activities. It is notable that in its urban renewal and area project the SDA has 'secured the active co-operation of the local authorities and others' (Industry Department for Scotland 1987, p. 89). This is well demonstrated in the GEAR project, where the joint committee of public-sector participants has helped to develop consensus and commitment. In later initiatives the SDA has used formal project agreements signed by the various participants, which set out agreed objectives, targets and timescales (Keating & Boyle 1986).

The LDDC shares with the SDA a system of formal and informal control under the direction of central government. It too is funded by a combination of central government grants and various self-financing measures, including land sales. It differs from the SDA in being a limited-life agency, set up for 10 years initially, although this may be extended. It has a much more streamlined organizational structure, with only 90 full-time staff and 88 fixed-term contract staff, and it makes much greater use of outside consultants. This mode of working is being adopted in an even more streamlined form in the second and third round of UDCs launched in 1987. The new UDCs are going to be limited to about 30 staff. They will be heavily dependent on local authorities to perform every-day processing tasks, such as the administration of planning and grant applications, and on private-sector consultants to provide development briefs and strategic policy reviews.

The LDDC has had a very poor relationship with the London boroughs in whose areas it operates. There have been repeated complaints from the boroughs that the Corporation fails to consult and in some instances fails to inform them of its plans and intentions (Docklands Consultative Committee 1985). The three boroughs have adopted very different strategies in dealing with the LDDC. Putting it crudely, it would appear that Tower Hamlets has gone along with its plans, Southwark has expressed defiant opposition and generally had nothing to do with the Corporation, and Newham has sought to negotiate and bargain. Yet each admits that it has had only a marginal impact on the Corporation's activities. By contrast, the early impression from the later rounds of UDCs is of a much greater effort at partnership with the local authorities, but it remains to be seen whether a genuinely co-operative relationship develops in the long term.

The institutional devices employed in the cases of Stockbridge Village and Coin Street are also very distinctive. Stockbridge Village Trust takes the form of a non-profit company. The chairman, Tom Baron, was appointed by the Secretary of State for the Environment. Other directors were nominated by the major participants, Abbey National Building Society, Barclays Bank and Knowsley District Council. In addition there are two community representatives, one nominated by Knowsley Parish Council and one directly elected by the tenants on the estate. However, none of these directors is officially there to represent their parent bodies or constituencies. Rather, they are appointed as directors of the Trust and must therefore act in the best interest of the Trust.

The company form of organization has been increasingly employed within local government, notably in the field of economic development. The best known examples are the enterprise boards set up by the GLC, West Midlands County Council and other authorities during the early 1980s. Other forms of public/private partnership organization have operated within local political structures. These include some 300 or so local enterprise agencies providing support for small businesses, the Groundwork Trusts which undertake environmental work, and local economic initiatives such as the Community of St Helens Trust and the Neath Partnership (Stoker 1988). The advantage of the company form is that it enables resources to be drawn in from both public and private sources, and it can act more quickly and in a more flexible manner than a typical local authority with its cumbersome committee procedures. The disadvantages of the company form are that public scrutiny of its decision-making and financial affairs can be severely limited. Meetings of the board are held in private, and an aura of confidentiality can surround all its activities. A high-handed style of decision-making can result, as well as a degree of slackness about financial accounting procedures. These weaknesses clearly feature in the case study of the Stockbridge Village Trust.

As the Coin Street example shows, effective popular planning involves a variety of community organizations. The Association of Waterloo Groups acted as an umbrella organization, maintaining a quasi-official image to present arguments to public inquiries and local authorities. The Coin Street Action Group engaged in more direct action, publicity and protest. The implementation phase saw the creation of the non-profit company, Coin Street Community Builders, primary and secondary housing co-operatives, and a new consultative body, the Coin Street Development Group. Community-based participation demands flexibility in institutional arrangements. A variety of community organizations, appropriate to particular roles and activities, is required to make popular planning a possibility.

The diversity of institutional arrangements for planning in the 1980s shows a trend away from local authority domination. The past decade has seen an increase in the number of semi-independent agencies in the planning field, and a corresponding decrease in the power and influence of local authorities. These new agencies shift some of the responsibility for local planning away from established local government structures, either to the lower level of neighbourhood-based organizations or to the higher level of more direct control by central government.

The neighbourhood-level organizations depend to a degree on local authorities relinquishing some of their planning powers. In popular planning, community organizations attempt to determine the direction of planning in a small part of a local authority area. The case study suggests that successful popular planning depends on the willingness of the authority to concede some of its power, and therefore on the degree of local control which the community can acquire. The private-management trust is also a neighbourhood-level planning agency, although not controlled by local residents. As the local authority is a necessary partner in the trust, it also depends on the authority conceding some power in the particular neighbourhood. Significantly, in the two case studies, both Lambeth and Knowsley conceded this power with some reluctance, under pressure from the GLC and central government respectively.

Quasi-governmental agencies are arms-length central government appointed and funded organizations. As such they represent the opposite, centralizing trend, which has also reduced local authority planning powers. The LDDC is directly responsible to the Secretary of State for the Environment and has effectively displaced the local authorities in planning its area. Although still formally responsible for planning policy, the role of the local authorities has been rendered ineffectual. Even though they have retained responsibility for most other local government functions in London's Docklands, the direction and pace of change has been dictated by the LDDC. In the case of GEAR, the formal role of the local authorities was more significant but, as the case study shows, GEAR was effectively led by the SDA which is directly accountable to the Scottish Office.

These changes in institutional arrangements can be seen as a response to the crisis in planning which we discussed in Chapter 1. Political opinion on both Left and Right has become increasingly sceptical about the efficacy of traditional local government structures and practices in planning, as in other fields. The Left has perhaps retained a stronger commitment to local authorities, reflected for example in the relatively central role which they play in public-investment planning compared with leverage planning. At the same time the 'New Urban Left' has favoured decentralization of local authority service provision, including in some cases support for popular planning (although popular planning can also appeal to right-wing and liberal political ideologies, for different reasons). Generally, the Thatcher government has attempted to remove a wide range of powers from local authorities, in favour of either market mechanisms or quasi-governmental agencies under its direct control. This strategy has been pursued in other fields, notably in the expanded role of the Manpower Services Commission in training and vocational education (Stoker 1988).

These moves to by-pass local authorities have also to be seen in electoral terms, with the 'New Urban Left' aiming to strengthen its base in local communities and the New Right attempting to weaken the powers of Labour-controlled local authorities and bring more Conservative voters to the inner cities.

While planning has not always been well coordinated under local authority domination, the emergence of new planning agencies at both the neighbourhood and the central government levels seem likely to produce a more chaotic pattern of development. All the 'new' styles of planning, that is every style except regulative planning, are to some degree in conflict with local government objectives and responsibilities. The more centralized styles conflict with local authority responsibilities for meeting a wide range of social needs in their areas, while the more localized styles come into conflict with authority-wide responsibilities. The already weak capacity of local planning authorities to perform any sort of coordinating role is further undermined by the institutional fragmentation of planning.

Politics and decision-making

Each of the planning styles we have considered has distinctive political characteristics and related forms of decision-making. The two local authority based styles, regulative and trend planning, are dominated by the actions of a relatively small group of councillors and officers. In most authorities the councillors and officers responsible for planning are given considerable discretion to interpret local policies and make decisions. Discussions with other departments and committees, and with other councillors in their role as ward representatives, take place when appropriate. But only in the case of very large-scale developments, with major consequences for the whole area, are planning committees and departments likely to experience direct interference from the Chief Executive, the political leader of the council, or the controlling party group. The main difference in the politics of these two styles is that regulative planning involves extensive debate among local professionals and politicians about market demand and strategic planning issues, while trend planning is much less concerned with these issues, preferring to trust the judgment of the market.

In Cambridge, at both county and district levels, professional planners and elected councillors argued long and hard about major planning issues such as the scale of growth, negotiated planning gains and environmental considerations. Conflicts occurred between planners and politicians, and among politicians on the basis of their party loyalties or local ward interests. In many respects leading councillors, with long experience of land-use planning decisions, became as expert as their professional advisors. The politics of regulative planning can here be seen to revolve around a technical–political axis. By contrast, in Colchester, where trend planning dominated, decision-making was more procedural in nature. The direction and scope of development were not seen as issues requiring the judgment of professional planners and councillors, since the market was the decision-maker and a generally pro-development attitude prevailed. In trend planning the planners still act as gatekeepers to the planning system, but, unlike the urban gatekeepers identified by Pahl (1975), their concerns are limited to non-strategic issues such as aesthetic control and conservation.

Both leverage planning and public-investment planning give leading roles to quasi-governmental agencies. Both have a corporatist political style, in that the effective implementation of policy is seen to depend on various agencies taking part in policy formulation. It is a characteristic of corporatism that state officials share some of their decision-making authority with selected external interests. These interests, in turn, recognize that some obligation is placed on them to act in accordance with the agreed policies because they have played a part in their formulation. In the case of leverage planning, a relatively pure form of corporatism exists in which the state agency seeks to involve powerful private-sector interests in order to ensure their co-operation and commitment. The LDDC discusses its planning frameworks and negotiates individual development projects with a variety of private developers, housebuilders, industrialists and financiers. With public-investment planning, on the other hand, a hybrid form of corporatism develops in which the interests invited to participate are other state agencies: we call this administrative corporatism. GEAR rests on the view that the involvement and commitment of a range of public-sector agencies is essential for an effective renewal scheme. In the process of policy formulation and implementation the joint committee provides a formal mechanism for inter-agency co-operation. This is supplemented by extensive and more informal communication among a range of officials.

The four planning styles discussed so far give only a limited role to the public and community groups. Our example of leverage planning, the LDDC, has at best a lukewarm attitude and at worst a positive hostility towards public involvement. Trend planning tends to limit the public's role to commenting on issues of detail and aesthetics. Both regulative and public-investment planning put more emphasis on public involvement, but while consultative mechanisms may be provided the dominant process is information-giving rather than genuine participation. The key decisions remain with elected representatives and officialdom.

The public also has a subordinate role in the case of private-management planning. In Stockbridge Village the key decisions were taken by a small group of the Trust's directors, backed up and supported by government and other officials. But the public has a very special place in private-management planning: they are there to be saved. We have described the Trust's relationship with local residents as paternalistic. Local people were encouraged to become more self-sufficient and were asked to 'share the vision' of the scheme's leading figures, more an invitation to be born again than to become involved in decision-making. Popular planning, by contrast, is premised on the very different view that the skills, knowledge and experience of local people can produce the best planning solutions. Politics and decision-making in this planning style takes the form of imperfect pluralism. The views of all kinds of local interests are sought in order to establish a broad consensus and to create an environment in which the wishes and concerns of local people can be understood and acted upon. The process is imperfect in the sense that, inevitably, some interests mobilize, while others do not. The Coin Street case study noted a shifting pattern of alliances, with community groups forming and dissolving and different types of political leadership in control of public authorities.

This review of politics and decision-making in planning shows that the traditional forms of local representative democratic politics survive only in the arena of regulative planning. Representative politics has in recent years gained a greater influence over this style of planning, with non-political planning committees led by officer advice increasingly giving way to more politicized committees that are heavily influenced by party political preferences. In other planning styles the local representative democratic mode, based around a technical–political axis, has been replaced by other forms of decision-making.

Other writers have remarked on this shift in the politics of planning, claiming to detect a general drift towards corporatism (Simmie 1981, 1985; Reade 1987). On the basis of the case studies, we would argue that the blanket use of the term 'corporatism' to encapsulate the direction of political change in planning obscures more than it reveals. Corporatism only exists in a relatively clear-cut form in the case of leverage planning. Public-investment planning adopts a hybrid form of corporatism based on consultation among state agencies. Trend planning involves negotiation with powerful private-sector interests over the details of development, but the central strategic decisions are left to the market and are not subject to corporatist bargaining. Similarly, private-management planning involves secret negotiations between public- and private-sector bodies, which might be described as corporatist. But these negotiations take place within the context of a commitment to 'saving' a run-down area, producing a political style which is dominated by paternalism. Popular planning is, of course, deliberately anti-corporatist, aiming to bring together a wide range of local interests in an open discussion of planning issues. The argument for a drift towards corporatism, while it has some credence, fails to capture the complexity of political change in the planning field. The fragmentation of planning in the 1980s has produced a corresponding diversity of political styles and forms of decision-making.

Conflicts and tensions

In the case studies we have described the particular conflicts and tensions which characterize each of the planning styles, and these are summarized in Table 9.1 under the headings of 'limiting factors' and 'principal interests to benefit'. By limiting factors we mean the major constraints on the planning style, which limit the extent to which it can achieve its own objectives. The legitimacy of each style rests partly on its effectiveness in its own terms, so that failure to deal with the limiting factors can generate tensions in the application of the planning style. Planning also has distributional consequences, since the spatial pattern of development and the resulting types of land use affect the accessibility of resources and facilities. From the case studies we can identify the principal interests to benefit from each of the six styles, whether these be owners or non-owners of property, locals or outsiders. Marked disparities in the distribution of benefits are a major source of conflict over planning issues, and this also affects the legitimacy of the planning styles. In the discussion that follows we look at each style in turn, focusing initially on the main beneficiaries, as well as the losers, and then turning to the limiting factors and problems of legitimation.

Regulative planning
Regulative planning can claim a wide range of potential beneficiaries, something which helps to explain its long period of consensus

167

political endorsement. As well as guiding the pattern of most new development, it has helped to resist unwanted change and maintain the quality of the environment in areas which would otherwise have come under strong market pressures, such as attractive suburbs and villages. Indeed, it is often said that the main achievement of postwar planning in Britain has been the prevention of urban sprawl. The main beneficiaries have been existing owners of land and property in attractive areas, particularly owner–occupiers and those who have access to this housing sector. Since regulative planning places limits on the availability of development sites, owners of such sites have also benefitted selectively when granted planning permission. Consequently, many large housebuilders with established land banks favour a highly regulated release of building land. These firms often make most of their profits from land speculation within a framework of restrictive planning policies (Rydin 1986).

Regulative planning can also claim to provide benefits for other sectors of the community. In a buoyant market it has often been possible to achieve negotiated planning gains from large development projects. Typically, these gains are features such as road improvements, public open space, improved landscaping and children's play areas, which can enhance the quality of the development. Although planning gains play an ideological role in helping to legitimate large-scale private sector development, they can also produce tangible benefits.

The losers in regulative planning are harder to identify. Principally, they are all those for whom market mechanisms do not provide, mainly in lower-income groups. People who cannot gain entry to the owner–occupied housing sector or take advantage of car-based suburban shopping developments can therefore lose out. To compensate for this, regulative planning has generally been accompanied by a large measure of public provision, for example in housing and transport. But another group of losers are those who, for example, could buy houses if more were provided. Restrictive planning policies have the effect of preventing market mechanisms from responding to many demands. The developers who would be willing to meet these demands must therefore also be classed as losers.

The ideology of regulative planning has therefore been concerned with balance and the 'general' interest of the community. Its legitimacy depends on achieving a broadly acceptable outcome, implying that political consensus is both desirable and possible. Public participation in planning was brought in to help to achieve consensus, but as we argued in Chapter 1, during the 1970s it frequently failed to do this and instead planning became more contentious. In every planning decision the appropriate balance of social gains and losses is contestable: What is the value of an area of open land as against the value of building houses? How much planning gain can a developer really afford to provide? Both general planning policies and the details of individual development control decisions are therefore continually open to challenge.

The legitimacy of regulative planning has also been questioned for its limited ability to achieve socially determined goals. Planning policies drawn up on the basis of elaborate studies of local needs and due consultation with interest groups still depend on the capacity of the system to regulate the private market. The main limiting factor therefore appears to be the strength of market demand. Without a relatively buoyant market regulative planning can exercise very little influence. Weak market demand has generated strong temptations to make concessions to developers, yet in the face of strong market demand the regulative powers of planning have proved weak. The planning system has therefore appeared to promise more than it can deliver, whatever the state of the market, and many sectors of the community have found it wanting.

Trend planning

Trend planning involves the lifting of restrictions on market actors, so its principal beneficiaries are those who can take advantage of new development opportunities. This means both developers themselves, including landowners and building companies, and potential users of new developments, such as house buyers and car-borne shoppers. The benefits of trend planning are distributed in a significantly different way from those of regulative planning. Whereas under regulative planning a few landowners stand to make large capital gains from occasional planning permissions, under the more permissive regime of trend planning a larger number of landowners can expect to receive planning permissions. Capital gains will therefore be more widely spread but generally lower, reflecting the increased supply of development sites. An increase in development will, in turn, attract more developers and eventually new residents to the area. It can therefore be argued that the benefits of trend planning go mainly to outside interests rather than to existing residents.

We noted above that large developers with established land banks generally favour regulative planning, which restricts the release of development land and keeps up prices. The volume housebuilders typically fall into this category. Other developers may seek less regulation, either because they do not have land banks or because they are unable to form a close enough relationship with local planners to rely on obtaining planning permissions. It is not always easy, however, to distinguish those developers pressing for deregulation from those seeking more land release within a firm regulative framework.

Advocates of trend planning also argue that the market is a better judge of consumers' wishes than are local planning authorities. Trend planning is therefore supposed to produce what people want, such as a better supply of new housing, which in turn is claimed to keep down the rate of house price inflation. These wider claims on behalf of trend planning are not very convincing, for two reasons. First, unrestricted private development will inevitably bring long-term disbenefits, namely the urban sprawl that was seen between the wars and that has, to a great extent, been contained since 1947. As well as eroding the countryside, sprawl creates higher than necessary infra-structure and transport costs. Trend planning has no means of coping with these unwanted 'side effects' or social costs of market processes. Secondly, the impact of a larger supply of new housing on prices is likely to be insignificant. New houses form only a small proportion of the total supply in most areas, and when prices do fall or fail to rise fast enough, housebuilders quickly reduce their output. Trend planning therefore has few claims to provide general social benefits.

The losers from trend planning are, by and large, equivalent to the beneficiaries of regulative planning. Existing owner–occupiers will see a deterioration in the quality of their environment, and potential development gains could fall for some existing landowners. Planning gain is, by definition, precluded in trend planning so that some social benefits are also foregone. This style of planning therefore threatens various established interests, including many which would otherwise naturally support the Conservative party. Its legitimation is rooted in the neo-liberal or New Right ideology of the Thatcher government, which generally favours the deregulation of private markets. The problem is that while some market or landowning interests will benefit from deregulation, others will lose out. In the past, the appeal of regulative planning to social consensus and the common interest has enabled existing property owners to protect their assets. These vested interests have been reluctant to convert to an alternative ideology of unfettered competition. In the long run they may take particular exception to being excluded from decision-making as the scope for public participation and locally determined planning policy is reduced. The main limiting factor in trend planning is therefore its capacity to retain some control over market processes, or at least the appearance of some control.

Popular planning
Popular planning is intended to produce direct benefits for a local community, in the sense of securing the kinds of development which

local people expressly desire. The beneficiaries could therefore be said to be the 'popular planners' themselves, although in practice those directly involved in planning are normally a small group of activists and community leaders. In principle, a popular plan should respond to the needs of all groups in the local community. The case study, like other examples, suggests that the ideology of popular planning favours lower-income groups and non-property owners. This reflects the fact that popular plans are often advanced as alternatives to commercial development, in defence of existing working-class communities.

Operating through a variety of community organizations, popular planning creates its own mechanisms for investigating local needs and consulting the community. At Coin Street this produced a range of consultative groups and working parties, as demanded by the various stages of the planning process. This method of planning can be effective but it may fail to uncover certain local needs or, perhaps more likely, it may become dominated by a particular local faction. It is therefore possible that the more organized and vocal elements of the local community will benefit most. Residents of an area with a popular plan might also gain at the expense of a neighbouring area, since the plan will demand the commitment of local authority resources. There can therefore be losers in popular planning, both within the immediate local community and in the local authority generally, and this may threaten its legitimacy. In the Coin Street case, few resources came from the boroughs but the area benefitted to an exceptional degree from GLC funding. However, it is hard to say who were the losers in the Waterloo community.

The main losers in popular planning are potential commercial developers and their clients, both essentially outside interests. Opponents of popular planning argue that providing benefits for the existing local community means foregoing more general benefits. For example, offices and high-class shops at Coin Street, and London City Airport at the Royal Docks, were both represented as contributions to the long-term commercial success of London. This leads in turn to a challenge to the legitimacy of the popular planning process, which is accused of only taking account of immediate local interests and excluding wider societal interests. These contrasting views of popular planning represent a clash of irreconcilable ideologies – local socialism versus free market development. The legitimacy of popular planning therefore depends on the acceptance of a socialist planning ideology in society at large.

The main limiting factor in popular planning is community control of resources, particularly at the implementation stage. Fighting for agreement on a popular plan is not enough to ensure that the plan will be implemented, as the Covent Garden story confirms (Anson 1987). The case study of Coin Street emphasizes the importance of retaining control over implementation. At this stage the plan is exposed to the pressures of resource constraints in the public sector and contradictory investment criteria in the private sector. This is the real test of a popular plan, and community control can help to prevent it from being appropriated or redirected by outside forces.

Leverage planning

Leverage planning aims to regenerate a market in land and property development through public-sector subsidies, infrastructure and site preparation. Insofar as this is a successful planning strategy, the immediate beneficiaries are landowners and developers. The renewal process will generate capital gains, as land values increase, and trading profits for builders and construction firms. Benefits will then pass to those groups who can gain access to the new developments, such as housebuyers and employees of any new industry. The distribution of benefits therefore depends on the extent and nature of the economic regeneration that occurs. The Docklands case is exceptional in the speed and scale of regeneration, but it does show what kinds of new development the market provides. These are principally owner-occupied housing and speculative offices, with some associated retailing and leisure facilities. The evidence suggests that few Docklands residents have yet been able to benefit from either the new housing or the new employment. In the short term, the main local benefits have come from improved transport facilities and possibly some of the retail and leisure developments, although there is evidence of increasing planning gains in the later developments.

It is argued by advocates of leverage planning that local residents will reap more benefits in the longer term. While few local people presently have the skills required by new employers, younger people will be able to train for office jobs and modern industry. This should raise local incomes and help secure access to new housing. The more successful the regeneration of the local economy, it is argued, the more local employment will be created, particularly in service industries. As we saw in Docklands, this case can be hard to refute when faced with something on the scale of Canary Wharf or the Royal Docks developments, but leverage planning will not always produce such spectacular employment growth. The other pioneering urban development corporation, on Merseyside, has been much less successful, and the other Enterprise Zones have brought few new jobs to their areas (Tym 1983). The main limiting factor in leverage planning is therefore the potential for a revival in local land and property markets, which will vary considerably between one city and another.

The uneven distribution of benefits means that leverage planning creates a major source of potential conflict between the existing community and new residents and workers, posing a challenge to its legitimacy. This is further reinforced by the organizational form of leverage planning, which by-passes the democratically elected local authority, and by its corporatist style of decision-making which largely excludes local interests. Potential conflict of this magnitude could only be contemplated by a government with a strongly ideological view of planning. Leverage is about overriding local interests and locally determined planning policies in favour of the presumed wider interests of the market economy, both in terms of the immediate gains to be had from development and construction and the longer-term benefits of employment growth. The regeneration of Docklands and other inner-city areas is thus part of a national economic strategy, as well as an innovation in local planning policy.

Another question about the legitimacy of leverage planning must concern the quality of the environment it is likely to produce. This was touched on above with reference to trend planning and the same point applies here, namely that leverage planning lacks the means to direct the form and content of development. Instead, it is based on accepting whatever the market will provide, which is a limited range of profitable land uses. In the case study, London Docklands was criticized for the lack of a truly public realm. The success of the LDDC has perhaps given it a greater influence over the quality of development in later schemes, but other areas will probably have to accept lower quality and an even more restricted range of development.

Public-investment planning

Public-investment planning is currently a style which aims to bring benefits to the existing community in a severely run-down area. In the past, this style of planning had a different role. Major examples of publicly financed development, such as the New Towns and central area redevelopment schemes, were quasi-commercial in nature, aimed to lead the direction of change, and were often heavily criticized for their impact on existing communities. In the 1980s public-investment planning has become a residual rather than a dominant style. As the case study of GEAR showed, it originated in the policy shift towards inner-city renewal in the late 1970s and is now seen as an exceptional use of public funds to deal with the most derelict areas.

The effectiveness of public-investment planning in creating benefits for existing communities is open to a number of questions. There is no doubt that public investment can provide resources and facilities that the private market does not provide – in housing, social facilities, environmental schemes and even employment – for the benefit of low-income groups. The GEAR project has produced social rented housing and low-cost owner-occupied housing, along with other facilities for local residents. But the population of derelict inner-city areas continues to decline. Even more significant is the continuing economic decline of these areas, to the extent that even public investment on a major scale cannot stem the tide. The benefits of this style of planning can look rather superficial in the face of high unemployment and persistent poverty.

Public-investment planning also poses the question of how far benefits are going to newcomers and outsiders rather than the existing community. Both housing and jobs may be taken up by outsiders, leading to a gradual change in the population structure. The anticipated growth in the population of GEAR resulting directly from the renewal programme, for example, suggests such an effect. However, this criticism can easily be overstated. Some change is inevitable and compared with the social change now occurring in London's Docklands, that happening in GEAR is negligible. In derelict areas, any benefits to the local community might be considered a worthwhile gain, even if a few others enjoy them too.

The legitimacy of public-investment planning has less to do with the distribution of benefits than with its longer-term viability. The GEAR case study suggests that a large commitment of public resources in a derelict area, comparable with that in London's Docklands, can lead to little or no regeneration of private markets. GEAR has produced a small amount of private housing but little new industry. The onus therefore remains with the public sector to manage the area for the forseeable future. This includes not only the maintenance of the renewed physical environment, but also a continuing responsibility for meeting social needs. With public investment in urban renewal being pushed into an increasing residual role, the future level of resources for projects like GEAR is in serious doubt. The potential losers from public-investment planning might, regrettably, include the local community of the future if sufficient resources are not committed to long-term management. This appears to be the main limiting factor for public-investment planning in the present climate.

Private-management planning

From the case study of Stockbridge Village one could easily conclude that nobody benefits from this style of planning, which has not yet been shown to achieve any significant measure of renewal. If it were successful, there ought to be benefits for developers, from housing renovation and construction, and for the occupiers of new and improved houses. The potential gains for developers are, however, severely limited by the fact that the market is in such a derelict state and unlikely to be revived. Participants in the private trust are in effect being asked by the government to accept lower than normal profits in order to bear a share of the social costs of renewal. If it works they can also expect to benefit from favourable publicity. But if potential gains for investors and developers are modest, the potential losses are not great either. Judging from the case study, the ideological importance of at least some appearance of success means that the private sector can expect to be bailed out in the last resort.

The local community can also expect some benefits from success-ful schemes, in the form of improved housing and social amenities. In fact, the improvement of half the dwellings on the former Cantril Farm estate has been the main achievement of the Stockbridge Village Trust, but there have been few other benefits. As with public-investment planning, some social change must be expected and some benefits will go to outsiders, but since the gains are few the question of equity hardly arises.

Private-management planning has little credibility as a means of renewing derelict areas. It is premised on the ability of the private sector to generate profits by efficient management and development, and to redistribute these profits in a paternalistic way so that lower-income groups can benefit. The case study suggests that, in the most run-down areas where regeneration is really needed, this simply cannot be achieved: the private sector is neither that effective nor that generous. In fact, to present even an appearance of success, large public subsidies have been injected. This in itself undermines the ideological case for free enterprise and threatens the legitimacy of this style of planning, since it contradicts the principle of renewal based on private management and the profit motive. When criticism of the organizational form and decision-making processes of this planning style is also taken into account, it is hard to see it succeeding as a mainstream approach to urban renewal. Its legitimacy seems to rest more on its conformity with the free-market ideology of the New Right than on its actual achievements, and its main limiting factor would appear to be its inability to move beyond tokenism.

175

10

Remaking planning:
conclusions and prospects

We have argued that the late 1970s and early 1980s saw the debate about planning fragmenting into advocacy of a number of distinct styles. The fragmentation arose out of a crisis in planning, a massive lack of confidence in the dominant approach of the postwar period. The debate about different planning styles has been part of the process by which a new dominant approach has emerged in the 1980s. An important part of this debate has been local experimentation with different styles, which we have explored through our six case studies. But this period of local experimentation is now drawing to a close. The prevailing political climate suggests that market-led styles will increasingly dominate planning policy. In this chapter, we consider the prospects for the future direction of planning policy in the 1990s and set out the problems and contradictions of the market-led approach, together with an agenda for an alternative debate which could challenge the new orthodoxy.

The dominance of market-led planning

Throughout the 1980s the six planning styles have existed side by side, but this state of pluralism is unlikely to last much longer. The Thatcher governments have consistently promoted market-led styles and attempted to undermine market-critical styles of planning. The Conservative election victory in 1987, which brought in a third Thatcher term, has sealed the fate of the market-critical approaches as far as central government support is concerned.

The remaining market-critical styles were survivors from the 1970s, where local communities or public-sector agencies had held onto positions of influence. Public-expenditure constraints have now killed off public-investment planning. There is not political will at central government level for popular planning. Those local councils that would foster it find themselves constrained by more

176

pressing financial problems and the impact of legislation promoting the contracting out of services, a major reform of education and new styles of housing provision. There is an element of support for regulative planning from conservationist and exclusionary planning elements among Conservative Party supporters, but these can be accommodated by a version of trend planning. The statutory planning system will therefore continue to shift along the spectrum of practice away from strong regulative planning towards the trend planning style.

Planning policy in the 1990s will therefore be characterized by the dominance of market-led styles, but it will also have learnt the lessons of the recent period of local experimentation with these styles. The new dominant approach will thus be a modified version of the three market-led styles that we have identified.

In more prosperous areas market forces will be given greater freedom by a land-use planning system reoriented partly towards their needs. Market criteria, already entrenched in development control decision-making, will become more significant at the plan-making stage. Planning for residential development has led the way with these changes. Joint housing studies provide the opportunity for housebuilders to become involved in detailed site allocations at the local level (Rydin 1988). A recent report by Coopers & Lybrand (1987) recommended the use of housing demand indicators in structure planning and the setting up of monitoring panels at county level, which would include housebuilders and estate agents.

However, just as the housebuilders have spearheaded the dilution of regulative planning, it is their statements which now indicate the limits of that dilution. The experience of the past decade suggests that a planning free-for-all is neither politically feasible nor commercially desirable. While seeking flexibility within a system of planning control, the housebuilders have come to recognize publicly the advantages of limiting market competition. The Housebuilders Federation have therefore criticized the government's Green Paper on development plans, arguing instead for a new tier of strong regional planning. These changed pressures from market actors, combined with the strong grass-roots support for planning controls which protect the amenities enjoyed by some groups, will shape the future form of trend planning.

'Flexibility' will be the keyword of the new style of trend planning. Development plans will be flexible planning frameworks. They will identify and publicize a set of preferred sites for development, thus helping to contain potentially damaging competition between developers. Even if the green belts remain sacrosanct in the face of development pressure, the allocation of development sites elsewhere will be responsive to market pressure and frequently amended in the light of changing market circumstances. Development control will implement the development plans in a flexible manner, and instruments such as the Simplified Planning Zones and the new Use Classes Order will create scope for the market rather than the planners to determine the details of development. For example, the new B1 use classification of mixed-use buildings could be anything from 100% office development to 100% light industry. To ensure that the planners remain aware of developers' needs in relation to the planning system, closer liaison will be established.

In less prosperous areas, where the market requires it, public subsidy will be used to lever in private development. While in practice the extent of leverage will reflect the degree of private-sector interest, the success of leverage planning in ideological terms will depend on the leverage ratios appearing to be very high. Recent experience has shown that such success can only be demonstrated in very specific local circumstances. Therefore the style is likely to be increasingly confined to areas where a small injection of public funds can stimulate substantial private interest. There will also be a growing emphasis on profits generated by rising land values and the scope for using those profits for further pump-priming. The effectiveness of the policy will be directly related to the level of private development profits and its applicability limited to the areas of greatest commercial potential. This is clearly the logic of the mini-UDCs that the government is proposing to declare in areas with only marginal problems; here a small amount of public investment could attract substantial private-sector interest.

In more depressed areas, attempts to lever in private investment are unlikely to be successful. Massaging the figures may improve the leverage ratios, but the resulting land uses will be very limited in scope and directed mainly at groups from outside the local area. The derelict areas will therefore be left to the fate of private-management planning. Yet the experiment in Stockbridge Village has shown that, without the political commitment to allocate public resources to such areas, private-management planning will be a charade. If resources are not committed, local planning will only exist at the level of rhetoric; if resources are committed, then the result will be leverage planning presented in the guise of the private sector taking responsibility. Through this process public resources will be allocated in an haphazard and wasteful way to shore up the pretence of a private sector capable of urban renewal.

All of these modified forms of market-led planning will be characterized by a responsiveness to market indicators. They will be capable only of meeting the needs of that section of society able to pay for goods and services. Those unable to muster the resources to indicate demands in the market will largely be ignored. The new built form will be limited to the most profitable types of development, with only a limited range of land uses passing the test of viability. The private-sector village, the out-of-town hypermarket, the science and business park, will change the shape of the built environment in line with the growing prosperity of selected social groups.

We have already seen this in the case studies. Trend planning tends to mean owner-occupied housing estates with limited social facilities or the latest in retail developments. In London's Docklands, leverage has produced mainly owner-occupied housing and speculative office developments. There is less scope for negotiated planning gain or indeed for public control of the form and content of development generally. The outcome is an environment which is only accessible to those social groups which can buy the housing and shop in the malls and hypermarkets. Where there are wider benefits, as in transport, these are normally provided by the residual public sector as a stimulus to private investment. Market-led styles of planning therefore produce a peculiarly distorted built environment consisting of a network of private realms, each one a closed and secure world, with no genuine public realm. The implications of this pattern of development are only slowly becoming apparent. The social groups which the market does not cater for are not only socially excluded but also spatially excluded. The market-led approach does not provide space in the built environment for those who cannot afford to pay for it.

This will also be a form of planning which is closely allied to a restructured economy and a new set of social priorities. It is well suited to accommodate and respond to rapid change. We have seen the way in which the outcomes of market-led planning favour new, emerging and in-migrating social groups over existing communities. Each of the market-led styles has the explicit aim of bringing in owner-occupiers, business and industry from outside the area. In buoyant market conditions, this is justified on the grounds of national economic growth and letting more people share in the benefits of attractive areas. In marginal and derelict market conditions, the newcomers are seen as a source of local regeneration. Since private markets produce profits by meeting new sources of demand, market-led styles of planning inevitably tend to favour emerging social groups in society over existing ones.

By contrast, market-critical styles seem stuck in a defensive rut, reactionary and traditional in nature. Regulative planning, particularly as it developed during the 1970s under the revised planning system, has tended to restrict new development which would worsen an existing environment or upset an established interest group. During this period county structure plans attempted to preserve villages and the countryside, and the conservation movement strove to prevent the destruction of historic towns. Popular planning has also been about protecting existing communities, with its roots in community struggles against property developers and urban renewal. Public investment planning, too, has come to address itself to the preservation of existing communities and their environment. All these styles of planning are based on the involvment of local populations in the formulation of local planning policy with relatively little influence from a regional or national context.

Market-led planning can be seen as a success in its own terms. It is well suited to aiding a modernization of British society according to the priorities of financial and property interests. It allows, indeed encourages, these interests to reshape communities and environments, and even to reshape the spatial pattern of economic activity. The precise spatial consequences of market-led planning are difficult to predict because of the footloose nature of development interest. While the local population is tied to a particular area, the finance underpinning much development has no such local allegiances. It is capable of moving into and pulling out of areas with a fair degree of speed if economic prospects change. The tendency is for such financial interests to congregate where profits are highest and surest. A growing disparity between localities in terms of private sector interest can only be the result if change in the built environment is left to market-led planning. The shift towards such styles is, therefore, closely linked to a more rapid spatial restructuring appropriate to a rapidly restructured national economy.

The associated political ideology stresses the inherent vitality and creativity of the private sector and the need for self-help rather than public sector support. However, the continuing abandonment of derelict areas by the same 'creative' and 'vital' private market will open up a space for a parallel set of values. Paternalism and the charitable zeal of groups dismayed by the apparently amoral attitude of market actors and the government will be focused on these abandoned areas. Community architecture is the current flagship of this moral movement in planning, led by Rod Hackney and supported by the Prince of Wales and Church of England leaders (Wates & Knevitt 1987). The religious overtones it carries, its involvement with Royal patronage and the ever present fear of urban unrest surrounding its presentation as an urban renewal policy, are all reminiscent of the 19th century. But, as Brindley (1987) argues, such a movement only manages to mask the massive withdrawal of public- and private-sector resources from these communities. Self-help and charity are of little aid to the victims of the spatial restructuring implicit in market-led planning.

Out of the processes of fragmentation and local experimentation, planning has been remade. The new approach combines deference to market forces with a package of limited planning controls and public subsidy. It focuses on areas where development is likely to be most profitable, and leaves less commercial locations to the rhetoric of the moral planning movement.

Contradictions of the new approach

While market-led policies are often presented, by the Left as well as the Right, as being perfectly in tune with the economic signs of the times, they are not an unproblematic tool of modernization. They contain many flaws and are open to a number of fundamental criticisms. First, a potential challenge to the dominance of market-led styles is clearly posed by the substantial section of the population not adequately catered for by the market. The poor, the unemployed, the isolated elderly, the disadvantaged sections of society, all have cause to oppose the new approach. Market-led styles of planning offer little to the large group of welfare dependents who lack any significant command over market resources. They cannot buy houses, and in relative terms they do not shop freely in the durable goods market and cannot pay to take part in leisure activities. The state might be expected to take care of those left outside the market place, but under the Thatcher governments it has shown less and less inclination to do so. Other groups, not directly adversely affected, may also feel uncomfortable with such socially divisive policies.

Secondly, a related point is that the new approach is associated with a weakening of the powers of local elected planning authorities and therefore of the status of locally determined planning policy. Market criteria have progressively been given more weight than local wishes in land-use planning. Where this shift has been opposed, central government has been increasingly willing to step in and take decisions over the local authorities' heads. With leverage planning and private-management planning, responsibility is taken almost entirely out of the hands of local authorities, who are left with a residual role. Overall, the accountability of planning decisions to local communities has been reduced. The need to override the concerns and fears of existing communities in order to facilitate change provides a strong pressure to centralize decision-making.

181

But market-led planning is not only lacking in local democratic control. It is marked by a very limited degree of any kind of accountability. The case studies of the LDDC and the Stockbridge Village Trust have emphasized the free-wheeling entrepreneurial attitude of the key decision-takers and the limited formal means of ensuring control over resource allocation. This lack of accountability even contrasts with the situation within private-sector companies and institutions, where very sophisticated control mechanisms are common. Even in the case of local-authority-operated trend planning, our case study indicated concern among opposition councillors at the practice of deciding policy behind closed doors. Where public funds are being made available to planning agencies, this lack of accountability has also begun to worry government departments, and attempts may be made to reinstate some control. The second round of UDCs declared in 1986/7 appear to be subject to tighter central control than the LDDC. This will undoubtedly hamper their ability to respond quickly to demands from developers for subsidy or investment and is, of course, contrary to the ideology of minimal bureaucratic control. Nevertheless, the weakening of mechanisms to ensure the accountability of planning agencies is inherent in the logic of market-led planning.

Thirdly, there are more generalized dangers implicit in the shift towards market-led planning, which could affect even those groups currently benefitting from Thatcherite policies. For these policies are based on the short-term maximization of profit which always motivates market forces. This 'short-termism' precludes any longer-term husbandry of natural resources or safeguarding of British productive potential. It involves the waste of resources both in terms of future generations' needs and, since it has a short memory as well as shortsighted vision, in terms of past investment in people and places.

Fourthly, the market-led approach carries within itself the potential for great instability, as the stock market crashes around the world in Autumn 1987 demonstrated. This instability is equally characteristic of the property investment sector motivated by short-term profit and subject to the new planning approach. The results of such instability are perhaps even more visible in the case of property investment owing to the relatively fixed nature of the built environment. When the inevitable risks, which are taken even by security-conscious investment agencies, result in losses rather than profits, the private-sector solution will be to withdraw. The 'mistakes' of the property market then show up as under-occupied or empty buildings and stand as a long-term reminder of the short-term commitment, footloose character and inherent instability of much private-sector investment. Yet the impact of these risks lies mainly in the future, and the Conservative government's policies have found little opposition on the basis of the future dangers they expose people to. Instead, opposition has been based around the current losers from the policies and local resistance to change.

The problems within the market-led approach to planning can therefore be characterized as its social divisiveness, its lack of accountability, it short-termism and its inherent instability. The next section goes on to outline the basis of an alternative debate on planning, which might suggest an alternative vision of the future.

An alternative debate

Those who oppose market-led approaches have been hampered in their search for an alternative policy stance by the lack of successful, concrete experiments in urban policy. The GLC's imaginative policy experiments were cut short by its abolition, and public-expenditure constraints have hamstrung other councils' attempts at local socialism. It is not our intention to fill this gap by setting out the ingredients of a new policy opposed to the dominant Conservative approach. Rather, we conclude by drawing attention to the issues around which the oppositional debate should focus. To present an already formulated policy would be to prejudge the outcome of that debate.

A first priority must be the restatement of the case for long-term strategic planning of the built environment. The critique of planning from the New Right, in terms of bureaucratic tangles, unresponsiveness and the stifling of private enterprise, has been too readily accepted. Benefits exist in terms of directing resources to areas of need and not just demand, protecting and conserving environments for the future, and preventing the excesses of speculative development booms. Such planning also contains the possibility of redressing the imbalance in economic fortunes between the different regions of the country. It creates the potential for working with existing communities rather than ignoring or replacing them. Attempts can be made to manage the adverse impacts of social change and minimize long-term risks. Underpinning the case for strategic planning is the positive intention to direct urban change in line with broader social goals rather than just following short-term development profits.

To be sustainable, such a view requires a much better knowledge of the ways in which interventions can be made in market processes in order to achieve planning goals. The existing tools of the planning system are weak in relation to market forces, and their impact is poorly understood. More powerful tools, such as the partial land nationalization measures of the Land Commission and the Community Land Act, have been crudely applied in the past, often in inappropriate circumstances. The interaction of market forces and public-sector policy in this area requires a set of strong yet finely tuned instruments of policy implementation. In particular, greater knowledge is needed of three land policy issues. What is the precise distributive impact of existing market processes? Where is the appropriate point in market processes to apply policy measures to deal with the adverse impacts? What is the likely success rate (and what are the likely side effects) of those measures?

It has to be recognized that strategic planning for change, even if not market-led change, may conflict with the inherent conservatism of local communities. Longer-term planning may result in proposals for local areas at odds with the wishes of existing residents. The views of neighbouring communities may conflict as well. It may prove impossible to meet the needs of a local community solely by urban change within their locality. Ways of reconciling the legitimate concerns of both the indigenous population of an area and the need for urban change as seen by those outside the area have yet to be fully explored.

Implicit in this issue is the need to rethink what local democracy means and how best to achieve it. Throughout our analysis we have frequently used the term 'community'. In doing so we have recognized the value of cooperation and fraternity that the term implies. During the 1960s the definition of 'community' was much debated and its widespread use devalued. Rather than resurrecting that academic debate, we would urge a flexible approach to self-defining local communities. Similarly, we would argue for an open-minded approach to the ways in which such communities can participate in decision-making. The experience of popular planning has thrown up some valuable lessons. Conventional representative democracy at the local level and local authority-led participation programmes may not always be the most appropriate means of involvement and certainly do not exhaust the democratic options. A range of organizational forms for representing local interests and dealing with competing claims should be explored. Aspects of communication within and between such organizations, and to those outside the organization, are also important in today's society.

But overshadowing all these issues of local democracy and effective strategic planning stands the difficulty of building alliances for a market-critical approach in current political and economic circumstances. The spatial effects of restructuring have contributed to this difficulty. The experience of urban change in particular

184

localities has created quite different social and political formations in different parts of the country. Glasgow, Sheffield, Liverpool and inner London have all suffered withdrawals of private investment and deindustrialization but their social structures are diverse, and among their political leaders there is not even a shared perception of problems, let alone solutions. Yet without a common platform at the national level on which to build positive policies for planning urban change, these and similar areas will continue to decline.

If market-critical approaches are to move beyond defensive campaigns then appeals must be made to communities in many different circumstances, often spatially separated. Without some mechanism for building on these diverse experiences the prospect is one of further isolated and sporadic oppositional struggles organized within historic and backward-looking frameworks. We need a powerful and coherent reformulation of the market-critical approach to unify and give momentum to these struggles. The alternative is the continuing dominance of market-led approaches, bringing in their wake enormous social inequalities, a massive waste of resources and chronic instability.

Postscript: planning in the 1990s

When we wrote the first edition of this book in the late 1980s, we were concerned to emphasize the fragmentation of planning, the way in which an apparently unified system of planning was being broken up under the impact of Thatcherism into a number of different, localized forms. These we termed 'styles' of planning to indicate their programmatic nature as much as their differentiated impact on the ground. We investigated these styles in practice through the medium of case studies. The case studies provided a snapshot of British planning at a particular time but they did more than just this: by capturing a point when the fragmentation was occurring they provided exemplars of the different styles as they were being 'remade'. Now, some 8 years later, these case studies provide a standard against which to measure the continuing remaking of planning and to discuss possible future directions of change.

We begin this Postscript with a discussion of the current context for planning. We then revisit our six styles and consider the relevant changes that have occurred over the last 8 years, both generally and in our case study areas where there have been significant developments. And finally, we consider the implications of our analysis for trends in planning and the emergent styles of the 1990s.

The changing context for planning: beyond the Thatcher decade

In hindsight the 1980s appears less as a distinctive period in its own right and more as a stage in an extended process of change. From the perspective of the mid-1990s we can see more clearly where that process of change was leading, with the prevailing sense that we have moved decisively into a new era. Some have termed this a postmodern era, others refer to late modernity or reflexive modernity (Beck, Giddens & Lash 1994, Giddens 1990, Harvey 1989). Without entering into the disputes over the terminology adopted and the extent to which there has been a radical break with the past, it is increasingly accepted that there is something distinctively different about current times. This arises from a nexus of new forms of organization for economic production, new patterns of social organization and new requirements for successful political activity. There is a new awareness of the importance of means of cultural communication and there is the increasingly pervasive influence of electronic technology in all fields, economic, social, political and cultural. At the same time, there is the new centrality of environmental concerns. This is clearly a broad ranging agenda for discussion. But here we can characterize the context for planning today in economic, political, social and also environmental terms.

(i) The new spatial economy

Economically, there has been a tremendous shift since the late 1970s which has produced distinctive forms of production and consumption. Commentators use a variety of terminology to describe the new forms of economic organization that we appear to be living with today: flexible specialization (Meegan 1988); flexible accumulation (Harvey 1989); or disorganized capitalism (Lash & Urry 1987). There is considerable dispute over the extent to which these concepts fully describe our economic systems, just suggest tendencies, or identify change in certain key sectors and areas rather than uniform changes. Briefly, these changes are associated with the increasing use of technology both within the workplace and for communication between workplaces. Many routine tasks can be replaced with such technology, leading to changed job content: some jobs disappear completely (the press compositor), others are deskilled (crafts within production), and others take on a broader range of functions (professionals doing their own typing). No longer is it necessary for economic efficiency that all the different parts of a manufacturing or service delivery process are included within one organization, let alone one site. The subcontracting of tasks outside the main organization is increasingly common with sophisticated information technology controlling the ordering and delivery of component elements.

This form of economic structure has particular spatial consequences (Ball & Pratt 1994, Massey & Allen 1988). Under modernist economic production there was a clear trend in favour of certain locations rather than others, based primarily on access to markets, labour and raw materials, and tight spatial agglomeration economies operated which magnified the local economic benefit of inward investment. However, under current patterns different parts of the economic activity will have quite specific requirements of their locations. Much subcontracting is dependent on a cheap and flexible labour force (for example component assembly); other subcontracting will be too highly specialized and require skilled labour (for example software programming). Some elements, such as research and development, need highly trained professional input and much attention has been paid to the quality of the local environment that is most likely to attract such professionals (Massey, Quintas & Wield 1992). But other elements will prefer underdeveloped locations, such as rural areas in developed countries or overseas locations, where low pay, absence of unions and part-time, flexible and often female workforces can be found (Massey 1984).

Therefore, economic enterprises are now using space actively as part of their organizational strategies, finding the most profitable location for each of the various elements that make up their overall economic activity. Footloose capital will spread these elements in a net, linked by communications technology, across regional, national and international space. Where possible, such capital will use the promise of inward investment or the threat of outward disinvestment to secure benefits from the state, financial or in-kind. This playing-off of states against each other can operate within all spatial scales, but for British planning it is the competition between local areas to attract footloose capital that is most important. Land use planning and urban policy in each local area becomes involved in the race to attract more enterprises than their neighbours and counterparts further afield. Ironically this is at a time when the multiplier effects of attracting such investment seem to be lessened since the communications net loosens reliance on local suppliers.

This is one of the factors that contributed to a fragmentation of planning styles across localities, along with the increased emphasis on the 'selling' of space, the marketing of cities and localities (Ashworth & Voogd 1990, Healey *et al.* 1992). And it is an emphasis that is increasingly fine-grained: it is specific localities rather than regions that are being sold. With the consolidation of this form of economic organization, the efforts of the local state in trying to manage the relationship between social, political, and environmental features of the locality and the needs of footloose capital, through the use of the land use planning system and urban policy more broadly, are likely to remain significant.

(ii) The new politics

Politically, the new economy has been associated with, and has perhaps necessitated, a new form of government, now widely termed 'governance'. Whereas 'government' used to refer to the formal institutions of decision making within the state, 'governance' is a wider term encompassing the inter-relation of governmental and non-governmental organizations. According to Rhodes (1995, pp. 1–2, original emphasis), 'governance signifies a change in the meaning of government, referring to a *new* process of governing; or a *changed* condition

189

of ordered *rule*; or the *new method* by which society is governed'. In the world of governance, formal hierarchies of central-local relations are less important than the complex patterns of cleavages created by numerous horizontal and vertical divisions. And these are divisions which cross over the traditional boundaries between state, economy and civil society.

There are multiple reasons why governance seems a more appropriate description of state activity in the 1990s. First, the acknowledged interdependence of the public and private sectors reflects the acceptance of limits to state action and, further, the desire to reduce the resource commitment and spending of government. This is as true of New Labour as it is of 'old Thatcher', in the guise of Majorism. Both main parties compete to be seen as committed to low taxation, and would appear to be in agreement with Bill Clinton that 'the era of big government is over'. Second, there are the fiscal stresses, the demands from the business sector and the changed expectations of the public as consumers, which arise from the economic changes outlined above. Third, the current political ideologies sit uneasily with older conceptions of the state. Both the remnants of New Right ideology and the emerging communitarian politics of New Labour have an anti-statist flavour and search for differently justified action by government institutions. Fourth, there is the empirical evidence of more and more actors searching for and creating networks, partnerships and new forms of alliance. The European Union in particular has been a galvanizing force in creating these new relationship between actors.

In very general terms, the various approaches to governance express uncertainty about the conceptual validity of sharp distinctions between market, state and civil society which were common in past work. Theorists with an interest in governance see the boundaries between sectors as blurred. Those studies with a political focus express a concern with 'a wide range of political governance mechanisms with no presumption that these are anchored primarily in the sovereign state' (Jessop 1995). The concept of governance directs attention to the distribution of power, both internal and external to the state, and the interdependence of governmental and non-governmental forces in meeting economic and social challenges. Governance is about a combination of governmental and non-governmental organizations working together. It is concerned with how the challenge of collective action is met; its focus is on the issues and tensions associated with this shift in the pattern of governing. Governance therefore represents a response to the challenge posed by a society that is prone to rapid change and characterized by complexity and diversity. Drawing on insights from cybernetics, the thrust of the argument is that a complex, diverse and dynamic social-political world requires forms of governing which are themselves dynamic, complex and diverse. New forms of governance are emerging that take the conditions of postmodern society seriously. Heavy-handed regulation would undermine the creativity and dynamism of a complex society. Thus governments require other tools of governance through which they participate in the conflict of forces by giving their support to certain actors. This subtle form of intervention involves identifying an area of interest, the forces in operation and the nature of the intervention that would produce a more desirable dominant coalition (Dunsire 1993).

Therefore local planning has to operate in a very different political world. The issues are no longer solely those of the relation of tiers of government, of national, regional, county and district plans. Rather, the challenge is to create networks and institutions of actors to forge collective action in a context of diffuse power. Partnerships become a more significant mechanism for generating and, at the same time, implementing policy; and this includes partnerships between public and private sectors, between governmental and quasi-governmental agencies, and again with non-governmental organizations, between the voluntary and business sectors, and so on. This generates new concerns about the meaning of democracy and brings us on to the new social context for planning.

(iii) Social fragmentation

Democracy used to mean formal representation and creating forums in which representative bodies, for example trades unions and elected politicians, could meet and debate. Nowadays the combination of tripartite corporatism (state, unions, capital) with electoral democracy is insufficient to meet the demands for involvement in public policy. Society itself is organized along much more fragmented lines, with multiple cleavages. This gives rise to myriad groups, each with a political voice. Planning therefore has to function within conditions of polyphony. The identification and legitimation of the public interest in these circumstances requires a different, more flexible form of planning.

The trend towards social and spatial differentiation and fragmentation is perhaps most evident on the broader European scale, with the disintegration of the Eastern European monolithic states and the collapse of the former Yugoslavian federation. The political instability these changes have produced has cast a shadow over the whole continent, through both local bloodshed and hardship and because, where such instability prevails, there is always the threat of a totalitarian response. The rise of fascist parties in Russia, Italy and elsewhere is evidence of this threat. Decentralization within nations also raises these concerns. It was within the decentralized local government structure of Tower Hamlets, with its mini-town halls, that the British National

Party was able briefly to capitalize upon the effects of the economic recession and withdrawal of welfare state provisions on certain sectors of the local population.

Yet many influential voices within planning and broader left-wing political circles have long called for decentralization as a way of bringing the state into a more democratic relationship with local communities and overcoming some of the bureaucratic tendencies of local government. It was also seen as a way of supporting political challenges to the policies of a Thatcherite central government, and more generally acting as a counterweight to the centralized state which, it was assumed, would tend to ally itself with the interests of capital and production, rather than those of the working class and consumption, particularly collective consumption needs (Boddy & Fudge 1984). More recently, the voices of green political theorists and activists have been added to calls for more decentralization (Dobson 1991, Eckersley 1992).

Such decentralization can be seen as a way of harnessing one of the more attractive features of postmodernism, its emphasis on polyphony. As a cultural phenomenon, postmodernity replaces a concern with broad, unified narratives which apply generally, with this emphasis on difference, on variety, on many stories told by many story-tellers. This, it is argued, could provide a space for hitherto under-recognized groups to find a political voice. Differences of gender, ethnicity, sexuality and locality become sources of newly heard accounts of experience and political demands (Heskin 1992). Of course, voices from the extreme right are also heard in this polyphony.

This raises the issue of how safeguards for variously defined minority groups, and for liberal, pluralist values in general, can be maintained in decentralized political systems. Some might argue that one should rely on local political struggles through local democratic means, both inside and outside the local state, to counter voices deemed unacceptable. Past working-class, anti-fascist and anti-racist struggle can be cited as examples of this political strategy. Others argue that the more centralized state can be used as a repository of such safeguards. Hence the move for a national Bill of Rights by Charter 88, the increasing use of the European Courts of Justice, and the resort to the United Nations to defend human rights.

Local state activity within land use and urban policy, engaged as it is in mediating local social relations through decisions on land use and urban change, will increasingly have to engage with the problems and opportunities raised by the polyphony of voices at the local level. This does not result in a simple conclusion on the benefits or otherwise of local decentralization but rather raises questions about the nature of the relationship between central and local states in promoting the space for democratic action by local groups and defending the rights of

minorities against the results of such action. This has long been a major
stated rationale for the planning system in terms of preventing middle
class NIMBYism ('Not In My Back Yard') from acting as a block on
development for working class (or any other) needs. And while this
rationale may be observed more in the rhetoric of planning than in its
outcomes (Evans 1993), its existence as a rhetorical device and a
political ideal is given added significance by the threats underlying the
new interaction of local voices. The role of the planning system in
legitimizing local political processes provides an opportunity for local
democracy, deficient though the exercise of that opportunity is in
current practice (for example Gyford 1995). The challenge is to create
a forum in which a range of voices can be heard and in which
there are opportunities for deliberation. It is not clear that existing
mechanisms of representative democracy at the local level provide an
adequate base for such activities and for extending citizen involvement
(Commission for Local Democracy 1995). But what is clear is that
planning cannot divorce itself from the changing social and political
world.

(iv) The new environmental agenda

The fourth dimension of change concerns the rising importance of
environmental issues, including global environmental issues, in driving
policy and the public agenda. Commentators on postmodernism have
come to recognize that environmental problems have a particular reson-
ance in this period. They see the growth of scientific evidence on
the environmental impacts of human activity, and environmentalists'
formulation of the impending environmental crisis, as an example of
how society is increasingly focused on uncertainty and risk (Giddens
1990). Many areas of society show this focus: the spread of share
owning and other paper assets and the linkage between the performance of such risky assets on market exchanges and individual households' wealth and life chances; the reduced permanence of job
opportunities and career structures in more flexible labour markets,
the so-called 'death of the job'; and greater instability in personal
lives, with more divorces and serial relationships, and diversity and
fragmentation in household patterns.

These changes in many areas of social and economic life are given
particular prominence by media institutions which regard such change
as the essence of 'news' and journalism. Thus, as commentators such as
Beck, Giddens and Lash emphasize (1994), it is the *perceptions* of
change, past and future, that are important as much actual change, and
this is equally the case for environmental change. It is our conceptualiz-
ation of the changed relationship between society and nature that is at
stake here. This is particularly important given the central role that

media institutions play in a postmodern society. Indeed, one of the defining characteristics of postmodernism is the dominant role of images, mainly created by the media, in defining social identities and social relations. Similarly, media coverage of environmental disasters and hazards, such as the oil spillage from the Exxon Valdez, skin cancer risks from the growing ozone hole, the newly confirmed greenhouse effect, or the spate of food-related health scares, has influenced the new environmental agenda which appears to be having at least a marginal effect on policy (Hansen 1993).

This is not to deny that there is indeed a real problem of economic activity reaching or even exceeding the capacity of environmental systems to cope with that activity. Rather, the prominent place that environmental concerns now have in political rhetoric, the existence of some state activity on environmental planning at different levels, and the stress associated with personal priorities and worries, are all a reflection of the interaction of personal experience, scientific data, media messages and the general structure of a society in which insta-bility, risk and uncertainty are commonplace. In these circumstances the predictive capacity of planning is in doubt. Risk and complexity is the norm and the dominant utilitarian calculus that guided planning in the past is no longer viable. New demands are being made of planning in these circumstances of uncertainty and complexity.

For the local state and the statutory land use planning system, the environmental agenda has provided a new problem and a new oppor-tunity (Agyeman & Evans, 1994). It has given the statutory planning system a renewed role in environmental protection. This role has several benefits for those engaged in planning activities. It is in keeping with the general focus of the planning system on the environment, though it involves a significant redefinition of what is meant by the 'environ-ment', from the local to the global, and from amenity to sustainability (Healey & Shaw 1994, Myerson & Rydin 1994). It is also in keeping with the public interest goals that planning has always avowedly sought to represent: protecting the environment and, in particular, the survival functions of the environment can readily be represented as in the public interest in a way that some of the economic development goals of the 1980s could not. Environmental planning also meshes neatly with the professional conception of a 'method' for planning, and the demand for environmental protection often involves a renewed call for strategic planning and a rediscovery of the synthesizing and integrative skills of planners (Rydin 1995).

Seen in this light, the call for enhanced environmental planning may appear as an attempted return to modernist planning practice. However, the emergence of the environmental agenda carries with it none of the certainty of the modernist period. Environmental planning has to deal as much with doubt as with certain evidence on the impacts at issue. Risk itself, including environmental risk, can never be planned away. Rather, judgements have to be made about 'acceptable' levels of risk, about trade-offs between costs and risk-reduction, even about the legitimacy of different assessments of the risks themselves, for example by Greenpeace and BNFL (British Nuclear Fuels Ltd). So planners dealing with the environmental agenda, which appears to offer new professional tasks, status and fulfilment, will also have to deal with them in the context of greater uncertainty and risk.

From six planning styles to two

In defining six styles of planning in the late 1980s we emphasized a marked political and ideological division, between those styles which were broadly market-led and those which were market-critical, and the relation of this division to local economic conditions and land and property markets. Since then, both the economic and the political contexts of planning have changed, and we have seen the growing influence of social polyphony and environmental awareness. In this section we return to our planning styles to ask if they are still valid today, and to consider the direction of planning in the closing years of the century. Some of the case studies were local experiments which remain highly relevant today, and in these cases we have updated them in some detail – for example, Coin Street. Others were over at the time of publication of the first edition of this book and merit little further comment. In other cases the local studies showed the implementation of national procedures and policies, for example in relation to trend and regulative planning. Here we will consider how the national scenario has changed and comment on how this is affecting local practice across different localities.

Our central argument is that planning styles in the 1990s show a narrower range of practice. Politically, there has been a withdrawal from 'extreme' positions, on both the left and the right, and an apparent convergence towards the centre. Planning operates generally in two sets of circumstances: on the one hand, the relatively buoyant market conditions of economic growth and new patterns of development in the changing spatial economy; and on the other hand, market conditions which we described as ranging from marginal to derelict, where the underlying problems are of decline and regeneration. Today we can see two dominant styles of planning reflecting these different circumstances, which we term 'responsive' planning and 'partnership' planning respectively. Each style encompasses a range of practice, as we will show, but in doing so it has narrowed the apparent range of political and ideological divisions. Responsive planning therefore

195

combines aspects of trend and regulative planning styles, in attempting to adapt to the new spatial economy; partnership planning combines aspects of leverage and popular planning styles, to bring about local regeneration. While underlying market conditions – growth or decline – appear to be the dominant factor, this is not the 'end of ideology' for planning. Postmodernity has brought a new political rhetoric and a new language of planning, but this is less likely to have healed the ideological cleavages of the past as to have, perhaps temporarily, obscured them in specific, local and pragmatic adjustments. Hence, we suggest that while the two dominant styles of planning show a relatively narrow range of practice, they nevertheless contain a wider range of potential divisions and future directions for planning ideology.

The two styles of planning that we do not expect to see more of in the next decade are private management planning and public sector investment planning, so these can be dealt with quite briefly. These styles represented the polar extremes of Thatcherite and 'old' Labour planning. Private sector management was intended to act as a showcase for the ability of the private sector to plan an area *on its own* and, further, to revitalize areas where the previous regime of state planning had failed. Yet as our case study of Stockbridge Village showed, the private sector was unable to fulfil these expectations. During the lifetime of the case study the experiment was acknowledged as effectively over and there have been no new developments here or elsewhere to suggest any future for this style. Rather, other experiments elsewhere have confirmed the style as an oxymoron. Handing an area over to the private sector cannot solve the problems created by prevailing patterns of private sector investment; the idea that the state wilfully destroys areas where otherwise there would be solid market interest in investment is a figment of Thatcherite imagination. Urban problems arising from public and private failures, usually in combination, require a degree of collective response that the private sector on its own cannot achieve. This was seen in Stockbridge Village itself, where the rescue package put together in 1986 confirmed the abandonment of a private sector-led strategy. A further financial restructuring of the project in 1990 confirmed this trend and secured the Trust's financial viability. An evaluation of Stockbridge Village by Liverpool John Moore's University (Evans & Russell 1993) found that many of the redevelopment and refurbishment plans had come to fruition. A survey also demonstrated considerable tenant satisfaction with the progress that had been made, and showed that the social mix on the estate had not changed as much as originally hoped by the scheme's promoters. After 1986, Stockbridge Village appears to have become a relatively successful example of partnership planning.

In a similar way it is now recognized that public sector investment on its own is an inappropriate response to urban problems in a democratic mixed economy. The result is costly and can be heavy handed in relation to the local communities. Even where sensitive management strategies have formed a coalition with those communities, the style faces problems of heavy demand on resources and a potential blindness to the workings of market processes in the locality. Public sector budget constraints now combine with an almost hegemonic view that the public sector benefits from working with the private sector; the result is that in practice partnerships of various types dominate the planning system. A return to a planning strategy *led* by substantial public sector investment seems unlikely. Rather, areas of public sector investment that persisted through the 1980s are now being penetrated by private finance. For example, roadbuilding remained a key example of public sector infrastructure leading private sector investment, development and land use decisions, but now the Private Finance Initiative requires public funds to work with private funds, effectively turning it into a form of partnership. The partnership nature of this initiative is made more apparent by the way it has worked in practice. The government's announced intention was that certain infrastructure projects would be funded solely by the private sector, but the financial reality is that these projects are rarely profitable in market terms and require a degree of public sector support. Indeed in some cases the level of support is so high that the private sector contribution seems almost cosmetic – the contribution from the owners of Canary Wharf to the Jubilee Line extension is one such example.

Responsive Planning

In the 1980s the distinction between trend planning and regulative planning was clear cut. We could see a marked contrast in the ways in which different local councils were using the mechanisms and procedures of the land use planning system. In regulative planning the direction of private sector development by the public sector was dominant; in trend planning the public sector's role was reduced to facilitating private sector development. The difference was who was seen to be in control of local planning and the goals to which the planning system – through development planning and development control – was being addressed. In the 1990s we see an accommodation between these extremes. A reshaping of local ideological maps has blunted the sharp contrast between local political goals for controlling and those for supporting the private sector. Local administrations have learned through their own experience that neither approach is particularly effective on its own. They have also learned, through examples of 'best practice' in other local areas, how to achieve more specific local goals such as providing facilities and conserving amenities. The ups and downs of local economies have made periodic shifts between the two styles necessary: trend planning always seemed more attractive when development was in short supply while regulative planning needed an active local development market to make it work.

We have now reached a relatively stable point along the spectrum from trend to regulative planning which we term 'responsive' planning – less controlling than regulative planning, more 'responsive' to market indicators, but not involving the abandonment of public policy goals to the single-minded drive to get more private sector development into an area. The local pressures towards an accommodation between the two more extreme styles have been reinforced by changes in policy and procedure coming from the national and international level. The Planning and Compensation Act 1991 was greeted by the planning profession as sealing the fate of trend planning and ending a period, which dated from Circular 22/80, in which local planning was required to give primacy to market pressures. Under this Act development control should have first reference to the local development plan; as previously, other material considerations, which may include market demand, should be taken into account, but the clear intention is that approved plans should be an indicator of where and how development in an area will occur. This goes alongside moves towards full coverage of all local authority areas by district-wide local plans or unitary development plans. The re-establishment of a plan-led local planning system moves us along the spectrum from trend planning towards responsive planning.

Another factor that has reinforced this shift is the increased importance of environmental policy in the 1990s. The rise of environmental issues as a political issue and social concern has been highlighted above. This has translated into a huge new area of policy and procedure. In part there has been a consolidation of the recent increase in domestic action on this front. Moves towards integrated pollution control, which was recommended by the Royal Commission on Environmental Pollution over a decade ago, were finally confirmed by the Environmental Protection Act 1990; this gave new responsibilities and duties to local authorities in relation to pollution control, matching those in relation to waste management. Similarly the Environment Act 1995 contains new provisions for dealing with contaminated land, an issue which has concerned central government for many years. But there is no doubt that much of the impetus for increased environmental planning activity has come from beyond the national arena, particularly from Europe and the United Nations. The European Union has turned out to be an extremely important force for enhancing environmental protection in Britain. A large number of measures have been taken by the European Commission and Parliament, leading to directives aimed at reducing pollution and increasing air and water quality. Environmental non-governmental organizations have found the European level to be a fruitful one for their lobbying activities. As a result many new measures operate at the local level, increasing the role of planning in regulating the environment. One particularly important measure is the introduction of environmental assessment of projects, expected to have a major environmental impact, under a 1985 European directive effective in Britain from 1988. The net result again has been a shift away from market-led trend planning towards the responsive style of local planning.

The importance of the new environmental agenda has been reinforced at the international level by the United Nations. Following the preparation and publication of the Brundtland report, *Our Common Future*, in 1987, the process has continued with the Rio Earth Summit, known more formally as the United Nations Conference on Environment and Development. Conventions on bio-diversity and climate change signed at that conference commit the British government to further action on environmental planning, much of which will impact at local level. There are proposals for local bio-diversity action plans and local air quality management strategies to be drawn up by local authorities. There has been an attempt to see development plans as a mechanism for achieving sustainable forms of urban development, particularly more energy-efficient urban forms (Rydin 1995). However, not all of this activity is simply a matter of local authorities using procedures to control and influence private development: much of it is focused on building local networks and meshing with the Local Agenda 21 process. This is more accurately seen as the partnership style of planning, which we discuss below.

While these pressures at local, national, European and international levels have prompted a shift away from the trend planning style, it has to be emphasized that 1990s responsive planning still differs from 1980s regulative planning. For example, it still bears the marks (some would say scars) of Thatcherism in having more limited goals for planning policy; marginal adjustments are seen as acceptable (mere 'mitigation' in the terminology of the new environmental planning); and planners have learned the lesson that market mechanisms have to provide the underlying impetus for development. In a sense the new buzzword of 'sustainable development' – the concept that drives the Brundtland and post-Brundtland process – encapsulates this view. There is the push for renewed planning activity, for a reassertion of the role of the public sector in achieving the 'common good' or 'public interest'; but there is also a recognition that economies have to be sustainable too, that needs are met by market mechanisms (at least within developed countries) and within current economic frameworks.

Partnership Planning

Just as trend planning and regulative planning have changed and merged to form a new style which now dominates planning in buoyant market conditions, so leverage planning and popular planning have tended to converge in current approaches to urban regeneration. The market-led concept of leverage has had to accommodate the demands of local communities and business interests for a say in the process, and some benefits from regeneration, while the market-critical approach of popular planning has had to work with, rather than against, a wide range of social and economic interests in order to achieve regeneration and development. The result is partnership planning, proactive and developmental, encompassing a wide spectrum of activities. At one end are top-down strategies in which the focus is on creating the conditions for profit-making. At the other end is the rise of schemes operating on a not-for-profit basis that involve charities, development trusts, community groups and voluntary sector bodies. Both of these strategies rest on combining public and private resources; both require institutional mechanisms and devices to create the conditions for partnership; and both involve a complex interplay of bargaining, negotiation and networking between actors in the public and private sectors.

As noted above, the new environmental agenda has also contributed to the growth of the partnership planning style. The prime policy focus here is the Local Agenda 21 process. Agenda 21 was a weighty document presented at the Rio Summit which set out a manifesto for achieving sustainable development. It was adopted by governments attending the summit and has resulted in the UN setting up a Commission for Sustainable Development and requiring national governments to prepare a strategy for sustainable development, which the British government did in 1994. However, it was also recognized that much of Agenda 21 could only be implemented at the local level and, as suggested earlier, local authorities and planners were very ready to take on the responsibility for a Local Agenda 21 process. Co-ordinated by the Local Government Management Board, this has taken the form of a number of local initiatives aimed at developing, implementing and – more importantly – raising support for local environmental improvements (Agyeman & Evans 1994). Networking has been a key feature of these initiatives, which have brought a wide range of actors from government, quasi-governmental agencies and the non-governmental sector (both corporate and voluntary) into the policy process. A recognition of mutual dependence, which is the hallmark of the partnership planning style, is also characteristic of the Local Agenda 21 process. What remains to be seen is how much change on the ground is achieved by what is undoubtedly an unprecedented amount of political interaction on environmental issues.

In order to explore the development of the partnership style, and in particular to illustrate how it has emerged from changes in the leverage and popular planning styles, we must now revisit the case studies of London Docklands and Coin Street to see how their respective styles of planning have evolved into different varieties of partnership planning. In each case, we take up the story in the late 1980s and follow events into the mid-1990s.

(i) London Docklands – the 'End Game'

When we looked at Docklands in the late 1980s, it was at the height of a dramatic nationwide property boom, in both commercial and housing development, and leverage planning was having a major impact on the area. The world's richest property development company, Olympia and York (O & Y), had just taken over the Canary Wharf project and there were even grander schemes awaiting the go-ahead in the Royal Docks. A third Thatcher government had been elected and the confrontational positions of the LDDC and the Docklands' boroughs were ceding to negotiation and compromise. But very soon the property boom had collapsed into a deep and prolonged recession, major developments in Docklands were put on hold, and even the mighty Olympia and York was toppled. By the early 1990s commentators were rushing forward with their verdicts on the LDDC experiment, eager to condemn it as a 'failure', a 'disaster', and a 'nightmare' (for example Ambrose 1994, Barnes 1990, Brownill 1990, Coupland 1992, Fainstein 1994). There was strong evidence to justify these views at the time, as the attempt to build regeneration on speculative property investment had failed so spectacularly. However, in the mid-1990s it is possible to detect a more positive view of Docklands, which is manifesting a degree of economic and social 'realism' previously excluded by entrenched conflicts. Docklands has changed permanently and is now anticipating the withdrawal of the LDDC, which has had to adapt its planning style to new conditions.

In the most comprehensive assessment to date of the impact of the LDDC, Sue Brownill (1990) identified three phases in the regeneration of the Docklands. The first two, the 'Enterprise Zone' phase (1981–1985), and the 'Big Bang' phase (1985–1987), which produced Canary Wharf, coincide with our account. The third phase followed the 'Black Monday' collapse of international financial markets in 1987 which precipitated the drawn out recession whose effects were still being felt well into the 1990s. With much of the speculative activity damped down, and the Conservatives still in power in Westminster, the LDDC and the Boroughs found a common interest in collaborating over the future of Docklands. There were two main features of this third phase of Docklands regeneration. The first was the injection of more public investment in the transport infrastructure to match the extraordinary, and unanticipated, surge of office building on the Isle of Dogs, a process which Brownill characterizes as 'reverse leverage', that is private sector investment stimulating further investment by the public sector. The main projects were increasing the capacity of the DLR, and a westward extension to Bank, opened in 1992 (an eastward extension to Beckton opened in 1994, and a further southern extension to Lewisham is planned); the Limehouse Link road, opened in May 1993 and considered the most expensive road ever built in Britain; and the plan to extend the Jubilee Line through Docklands to Stratford. Olympia and York promised substantial contributions to both railway projects. The second was the introduction of a social component to the LDDC's activities, following the Memorandum of Agreement with Newham, of September 1987. The Corporation struck 'planning gain' deals to extract some social benefits from planning permissions. More significantly, it set up a Community Services Division and introduced social objectives into its 1988 Corporate Plan. As a result of this the LDDC has provided grants to support community projects and established community trusts, and in 1989 it introduced a social housing strategy. While this was criticized as little more than public relations, it represented a significant departure from the LDDC's original brief and produced concrete benefits – for example, by 1995 about 8% of the Corporation's total capital expenditure had been spent on social housing (LDDC 1995).

Brownill's evaluation of the Docklands experiment as 'another Great Planning Disaster' was subsequently underlined by the collapse of Olympia and York part way through the construction of Canary Wharf, in May 1992. Olympia and York had convinced the property world, and a large number of banks, that they could ride out the recession and successfully complete this project. Their failure had far-reaching repercussions for Docklands, for leverage planning and for international property finance. Fainstein (1994) identified four specific causes of the collapse of O & Y: the financial situation of the company, which found all of its investments slumping at the same time; the personal characteristics of the owners, the Reichman brothers, who had an exceptionally high tolerance of risk; government policy for Docklands regeneration, which relied too heavily on the private sector to achieve 'public ends'; and the state of the international commercial property market. Yet in spite of these specific conditions, she saw this collapse as a representative event which 'expose[d] the fatal weakness of relying

heavily on property development to stimulate regeneration', namely ⌉ ⌐
the inevitability of speculative over-supply.

The Docklands regeneration has now moved into its fourth and
final phase, the 'End Game'. This phase began in 1994 with the
forming of Canary Wharf Ltd to take the failed O & Y project out of
receivership, and the announcement of the 'de-designation' of the
LDDC and its progressive withdrawal from the area, with an end
date of 31 March 1998. This stage is characterized by a process of
'accommodation' – of the boroughs and the LDDC to each other's
roles and interests, and of local residents to the changes that have taken
place, whether or not these are still defined as a 'disaster'. There are
four main aspects to the last years of the LDDC: completing the
infrastructure; the partial recovery of the property market; the exit
negotiations; and the anticipated end of 'Docklands' itself.

Infrastructure has proved to be the deciding factor in the long-term
viability of Docklands as a commercial, and probably as a residential,
location. On the principle of leverage, commercial projects like Canary
Wharf were meant to pay for much of their own essential infrastructure
and at one point it looked as if this might really happen – O & Y
promised to contribute half of the £800 million estimated cost of the
Jubilee Line extension. But any hope of this major 'success' for leverage
disappeared in the property market slump and the collapse of O & Y.
The Jubilee Line has subsequently proved an embarrassment for the ⌉
Government's Private Finance Initiative, as the cost has risen to at least
£2 billion and the private contribution, still only £400 million, has
had to be supported by the European Investment Bank. The project
was finally given the go-ahead in October 1993. The total bill for the
LDDC experiment will prove extremely high: the Corporation itself
is expected to have spent at least £2 billion by 1998; the Jubilee line
will cost at least £1.6 billion; and a further £1.35 billion will have
been spent on roads and the DLR (*Independent* 1995). With a total
government contribution of over £5 billion, leverage has turned out
to be far less successful financially than it appeared in the late 1980s.
Set against a claimed private sector investment of £6.1 billion by 1995
(LDDC 1995), the 'leverage ratio' will turn out little better than 1:1.
It is notable that the government has continued to increase spending
in Docklands while it has been cutting expenditure in almost every
other sector. This suggests that there has been a political commitment
to the 'success' of Docklands which has outweighed its real benefits,
especially when compared with the needs for investment in other areas
or projects. Nevertheless, it is the commitments to infrastructure in
the 1990s that have 'rescued' Docklands from stagnation and set it
on the road to a more stable future.

The property market has made a slow recovery to 'realism' from the profound recession which followed the wild boom conditions of the 1980s. The housing market has stabilized with an increase in private lettings and the acquisition of unsold properties by the LDDC and housing associations. The commercial property market has also revived with Canary Wharf largely let – the Mirror Group and other news-papers have turned 1 Canada Place into 'Fleet Street in the air' – and in the financial sector the significant arrival of BZW in 1995. The recovery is partly in anticipation of the opening of the Jubilee Line in 1998, but the LDDC's vision of Docklands as the 'third node' of London's principal office market, after the City and the West End, still seems over-optimistic. Instead, it is likely to become established as an important secondary office market, more like Croydon or Hammersmith.

The imminent demise of the LDDC has led it into 'exit nego-tiations', building on the accommodation with the boroughs which it has pursued in recent years. Packages of social benefits have been agreed – a school, a district centre and a community trust fund – which are being represented as the LDDC's parting gifts. Perhaps the most important aspect of the Corporation's 'exit', however, is its role in helping to shape the borough's Unitary Development Plans. This process is helping to bring to an end the coherence of the Docklands as an identifiable place. Docklands was the product of an economic geography of the nineteenth century, its physical isolation reinforcing its social and cultural identity. A diverse new population has now moved into the Docklands: the inflow of high-income owner-occupiers, and some local house buyers, and low-income council tenants from other parts of the boroughs, has almost doubled the 1981 population. Social and racial tensions have grown as the homogeneity and social solidarity of the area has broken down and the long-established residents have seen their job and housing opportunities decline (Docklands Forum 1993). The 1970s regeneration strategy, the LDSP, with the creation of the Docklands Joint Committee, tried to hold on to the concept of the Docklands as a distinctive place, while proposing its reintegration with the surrounding boroughs. The LDDC redefined Docklands' identity again, but as a marketing concept rather than a place, in which it had neither an historic nor a long-term future interest. Ironically, the successful marketing of development sites with no overall strategy for the area has led to the fragmentation of Docklands, which is now 'less a place than a series of subcentres of economic activity' (Edwards 1992), a 'collage city' which stands as a classic example of postmodern urban development (Harvey 1989). The LDDC has played a part in 'regenerating' Docklands, but it has also overseen its restructuring and ultimately its break-up as a meaningful entity. The handover to the boroughs will further weaken the identity of Docklands, whose name will fade quietly into the past.

The style of planning in Docklands has changed markedly since the late 1980s, and the experience of the LDDC has led to major changes in the concept of leverage as an approach to planning. Leverage planning was used by the LDDC to bring about the 'regeneration' of the Docklands, with no regard for the local impact of the types of development that were 'levered in'. It inevitably produced a conflict with the boroughs which had specifically sought to revive their local economies and social infrastructures. Leverage planning therefore had to be imposed from the centre and address itself to opportunist speculators. In its own terms, it was initially successful in bringing about new housing and office development, but the speculative property market crashed and leverage failed the critical test of attracting private investment in even essential infrastructure, let alone in development to meet social needs. From the end of the 1980s leverage planning began to be redefined. The activities of the LDDC were moderated to bring it into line with shifts in urban regeneration policy generally, to include wider consultation and explicit social objectives. Later UDCs were set up with more local involvement and consultation from the start, and subsequent policy developments like City Challenge and the Single Regeneration Budget have moved away from pure leverage to various forms of partnership. Local authorities, voluntary organizations and businesses have been brought together to work out joint projects and funding bids, with a requirement for wide local consultation. The LDDC has not become a partnership, but it has moved in this direction as the need for a more balanced and consensual regeneration of the Docklands has been acknowledged. The Wild West capitalism of the 1980s has been toned down in favour of a more sensitive management of regeneration and economic restructuring. Public investment is still expected to stimulate private investment in urban regeneration, but in place of a blanket policy of leverage there are a host of specific joint projects, which attract wide-ranging support, are competitive and represent value for money. The free-market theory of leverage planning has therefore been displaced by the consensual approach of partnership planning.

(ii) Coin Street and the South Bank

Popular planning emerged from the activities of oppositional groups fighting for local interests, principally against the threat of commercial development but also against the imposition of insensitive or ill-considered projects by the central or local state. It had no formal role, but operated in the interstices of the planning and development system where it had to define its own role and functions. These changed in response to the outcomes of campaigns – mostly lost but a few won – and to new opportunities. Coin Street was a standard bearer of the popular planning movement largely because it moved on from confrontation and campaigning to become a new type of entity, a community developer, with the remit of implementing a popular plan. Since the late 1980s, Coin Street Community Builders (CSCB) have been working out what a community developer can do and how it can remain true to its popular roots. In the process, they have moved away from radicalism and forged common interests with a range of commercial, governmental and quasi-governmental organizations. While they continue to pursue popular planning goals, their new activities have extended the range of partnership planning.

By the end of the 1980s Coin Street Community Builders were a well-established presence on the South Bank, with all of their initial developments completed. The first cooperative housing scheme, Mulberry, was completed in March 1988. The temporary market on Gabriel's Wharf also opened in the Spring of 1988 with craft workshop and retail units, a garden centre, cafés and restaurants. The conversion scheme for Stamford Wharf had been chosen, including cafés, bars, a rooftop restaurant, and retail and workshop units, with 80 cooperative flats on the upper floors. They continued to pursue the objectives of the popular plan in the 1990s, with the completion of a second cooperative housing project, on Broadwall, to an award-winning design by architects Lifschutz Davison. The same architects took on the complex Stamford Wharf conversion, which became a prominent symbol of the regeneration of the area. A change of name to Oxo Tower Wharf, based on its landmark tower, emphasized the central place of this building in CSCB's popular image. At the time of writing, Oxo Tower Wharf had its first residents in cooperative flats, and the lower floors were due to open in 1996. CSCB was left with three major sites for future development, based on their original planning consent: the remaining large car park, scheduled for housing; the site of the Coin Street Design Centre, for workspaces; and Gabriel's Wharf, eventually intended for housing. It therefore saw its role as community developer continuing for some years.

However, the South Bank and Waterloo were changing around CSCB. While the defeat of Greycoats in 1984 had kept the property developers at bay for a few years, by 1989 several new developments were being planned for the South Bank which would have an impact on the Coin Street area. These began with the new terminal for Channel Tunnel trains at Waterloo Station, which opened in 1994; a major commercial development by P & O on York Road; the conversion of County Hall into a family hotel by its new Japanese owners, Shirayama; and improvements to the external spaces of the unpopular

South Bank arts complex, based on a dramatic 'crystal palace' design by Sir Richard Rogers. Other schemes came along, including the Jubilee Line Extension (approved in 1993), with two stations in the area; and the British Film Institute (BFI) plan for an IMAX cinema on the 'bullring' site at the southern end of Waterloo Bridge. To the west of the area, there were new plans to develop the Effra site at Vauxhall Bridge; and to the east the Tate Gallery's proposal to convert Bankside Power Station into a Museum of Modern Art, with the Globe Theatre nearing completion in the same area. In the early 1990s, the South Bank was no longer a backwater but a central focus for the regeneration of London's urban environment.

Therefore, at the start of the 1990s, Coin Street entered a new phase in which CSCB's plans for social and public uses on the South Bank faced new types of threat from new quarters. Their response illustrates many of the significant trends in planning styles which have emerged in the first half of this decade. There are two main points to note: first, the development of a broader role for the organization than simply that of land owner and developer of specific projects; and second, related to this changing role, the emergence of tensions between CSCB and other community organizations in the Waterloo area. The changing role of the community builders is seen most clearly in the establishment, on their initiative, of the South Bank Employers' Group (SBEG). At first this was a relatively informal group of organizations based at the South Bank who recognized their common interest in the quality of the urban environment. The initial members were Shell International, The South Bank Centre (i.e. the arts complex), London Weekend Television, the British Film Institute, the Royal National Theatre, IBM, HM Customs and Excise (who had moved into Sea Containers House in 1988), IPC Magazines and J Sainsbury. Later they were joined by the various railway companies operating out of Waterloo International and several other firms.

SBEG began by commissioning studies of the South Bank area from planning and urban design consultants. They progressed from studies to proposals, including a new piazza in front of the Victory Arch at Waterloo Station, a 'spine route' along Upper Ground, and extensive landscaping, signage and street furniture. They bid for government funding for four projects through the Single Regeneration Budget (SRB) (at the time of writing, a start is due to be made on the Upper Ground improvements early in 1996). SBEG also joined with central and local government bodies to form a public/private sector partnership for the South Bank. With this increase in support, from both local organizations and the government, SBEG put itself on a more permanent footing with the appointment in 1995 of a chief executive and an administrator, based at CSCB's Upper Ground offices.

The changed role and interests of CSCB started to create tensions in its relations with other community organizations, notably the WCDG. An incident involving a proposal by LWT to build an office block on its Prince's Wharf site, leased from CSCB, caused a rift between the two groups. By initially supporting the scheme, in exchange for a 'planning gain' deal for community facilities on another site, CSCB appeared to have become pro-offices. WCDG saw this as a betrayal of the popular plan's central principle of opposition to office development at Coin Street. In the event the scheme collapsed but it had affected the perception of CSCB in parts of the local community. In an interview with *Time Out* in 1994, Margaret Mellor of WCDG said that the South Bank had 'turned its back on the rest of Waterloo', and that CSCB 'has become like any other developer'. Not all of the current wave of new developments were accepted by local residents either, particularly the speculative or commercial projects which seemed to have little relevance to local needs. The IMAX cinema, for example, although welcomed by CSCB and Mulberry Coop, and granted planning permission by Lambeth, was condemned by WCDG as an 'eyesore' and heavily criticized by other groups for its likely impact on rough sleepers under Waterloo Bridge.

As the South Bank became the focus of all kinds of development interests, the planning process was changing around it. Southwark and Lambeth prepared their Unitary Development Plans, and plans for Coin Street remained essentially unchanged from the popular plan. However, local MPs and councillors saw the need to broaden consultation on the increasing proposals for change on the South Bank and in October 1994 the South Bank Forum was established. Planning was proceeding on the basis of much wider consultation between the various development agencies, public and private, political representatives and potential 'users'. At the same time, many of the agencies were attempting to work together in a variety of partnership arrangements. As well as SBEG, other partnerships included a joint bid for SRB funding by Lambeth, Southwark and the local TEC; a Cross-River Partnership comprising Lambeth, Southwark, Westminster and the Corporation of London, which also put in SRB bids; and the all-embracing South Bank Partnership, with representation from central government (the Government Office for London), local government (Lambeth and Southwark), elected representatives (ward councillors and local MPs), and local business (SBEG).

The events of the early 1990s show that there has been a marked shift in the style of planning at Coin Street, as the popular plan of the mid-1980s has been gradually implemented and new development pressures have emerged on the South Bank. When the local community won its famous 'victory' over the speculators, it represented an island

208

of regeneration in a relatively stagnant area. By the mid-1990s, the South Bank was the site of a host of new projects and improvement schemes, involving the public and private sectors, separately and together, and Coin Street had been redefined as both a model development and a catalyst for wider change. This seems to reflect a shift from a modernist concept of regeneration, based on planned provision for measurable local need – the GLC's 'community areas' concept – to the essentially postmodernist concept of 'cultural regeneration', where arts and culture provide a focus for change and renewal (Bianchini & Parkinson 1993).

In this context, two components of Coin Street stand out for particular note. The first is the gradual redefinition of building types from the original (unbuilt) schemes of the 1970s to the award-winning architectural projects of the 1990s. The original schemes were essentially conventional council housing, designed according to the prevailing models of low-rise, high density (Scoffham 1982), and this resulted in the Mulberry Coop scheme. The 1990s schemes, in marked contrast, are manifestations of an entirely different urban architectural language. The Broadwall housing employs striking contemporary forms, including an ironic 'tower block', and is intentionally open to the surrounding area in opposition to the defensive enclosure of the Mulberry scheme. Oxo Tower Wharf includes apartments on the upper floors which, although relatively conventional, are redolent of the 'loft living' (Zukin 1982) trend which is spreading rapidly across London. This conversion provides a complex mix of uses and users, in the manner of the anti-modernist city championed by Jane Jacobs (1961), Leon Krier (1978) and others. Mixed use is a way of undoing the land-use segregation which modernist planning brought about and which helped to destroy the pre-existing urban vitality. At Coin Street, it is intended to bring 'life' back to the South Bank and the Waterloo neighbourhood, with a varied mix of working, living, shopping, eating, relaxing, and a diversity of cultural activities.

The second postmodernist trend seen at Coin Street is the explosion of design and spectacle, and the emphasis on image and appearance. The mundane architecture of the GLC and the provincial architects who began the Stamford Wharf conversion was exchanged at the earliest opportunity for the eye-catching style of Lifschutz Davison. As well as attracting the attention of cultural critics and consumers (Coin Street began to be featured in *Time Out* and visited by restaurant critics), this gave CSCB a credible basis on which to engage in debate with other operators on the South Bank. Gabriel's Wharf, as well as being a focus for new, 'cultural' uses in the area, became the site of the successful annual Coin Street Festival from 1990. All of this gives the message that Coin Street is as much about design, culture and the arts as any other South Bank organization, and that this is now leading the direction of regeneration. Those who do not wish to go down this road, and this would appear to include the WCDG, are left behind in the dust of modernism and its old-fashioned concepts and values.

Coin Street illustrates a further trend of the 1990s, which is the fragmentation of planning amongst many agents, combined with a consultative, consensus-building approach which attempts to lock these agents into pragmatic partnerships to achieve particular, local goals. The community builders have spawned several agencies, including Coin Street Secondary (CSS), the 'mother' housing coop, and Coin Street Management Services, which manages the open space, to become a partnership themselves. Also they have established the wider partnership with local employers, SBEG. Both CSCB and SBEG have joined other partnerships to carry out studies and reviews, and to bid for regeneration funding, involving government, quasi-government and non-government agencies, as well as commercial interests. This is a manifestation of a decentralized, polyphonous style of planning in which the voices of the 'community' have become some amongst many, rather than a unified and dominant voice in a specific locale. The old language of popular planning, with its references to 'community struggle' and 'community victory', is now an outmoded rhetorical style. Regeneration at Coin Street is no longer a zero-sum game, with clear winners and losers, but something more akin to a permanent revolution, with constant opportunities for the various players to advance and change. While some will lose out in this new forum, those who fail to engage in the debate are unlikely to advance beyond 'go'. The question of accountability, which is inevitably raised by this competitive and disjointed approach to planning, has also fallen off the agenda: CSCB now asks to be judged not by to whom it is 'answerable' but by its contribution to the regeneration of a central London neighbourhood. This will be partly in terms of housing, jobs and amenities – the conventional planning criteria – but also in terms of how it has helped to redefine the experience of urban living in this part of the city.

Popular planning has lost its radical edge at the same time as popular opinion has been drawn into the planning of urban regeneration. The populist dimension of planning is now central to the partnership style, in which a greater number of community groups are consulted or work as full partners in regeneration projects and funding bids. The oppositional politics of planning still has a vigorous life, but it remains on the ground it started from, resisting road building or environmental destruction by direct action and lobbying. At one level, it could be argued that the popular planners have been incorporated in the formal processes of planning, while the true radicals remain outside and against the system. However, this would be to ignore the wider changes in planning, which has moved away from a strictly formal and procedural approach to a more open and flexible approach which responds to many influences. Popular voices have more chance of being heard in this style of planning than they did in the past, and community groups have found that they can influence outcomes, at least at the local level. To protest and resist is to define oneself outside the scope of partnership, but to develop popular plans and to work with other agencies gives at least some chance of success.

Returning to these two case studies has demonstrated just how much planning has changed in the 1990s in the context of urban regeneration. Coin Street and London Docklands now illustrate variants of the partnership planning style. They still show marked contrasts: the Coin Street builders continue to pursue a 'bottom-up' approach with a commitment to non-profit development, but they are now linked in with central government policies and commercial objectives; in Docklands planning still rests on a dominant 'top-down' approach and development for profit, but with strong connections to locally derived, non-profit plans and projects. Where they are similar is in the movement to centre stage of partnership structures for planning and implementation of change, effectively defining a new dominant planning style. Partnership, which has been practised in various forms for many years (Brindley & Stoker 1988), has come to the fore in the 1990s as the dominant style of urban regeneration. As Bailey and others (1995) argue, this was partly in response to the lack of consensus and the many contradictions which afflicted the government's preferred market-led approach. But as our case studies demonstrate, it was also a pragmatic adjustment to the exigencies of implementing urban regeneration plans in changed economic and political conditions.

Conclusion

The idea of distinctive planning styles remains valid. It captures developments in the practice and intentions of actors notwithstanding any formal change in the legislative framework for planning. We also restate our commitment to case studies of planning in specific locales, which both capture the reality of local planning in those areas and at those times, and further help us to explore the impact of change. The fragmentation of planning in the 1980s, however, has given way to the dominance in the 1990s and beyond of two planning styles: responsive planning and partnership planning. In this brief conclusion we consider the task facing the planners who are implementing the current styles of planning, and how their professional roles can adapt to the new conditions of wholesale change, uncertainty and risk.

211

POSTSCRIPT

The period into the twenty-first century will provide a distinctive context for the operation of the planning system. Managing land use and urban change to generate and encourage economic growth will take place within a highly specialized and spatially diverse economic system. At the same time, the planning system's historic responsibility to political pluralism and public participation will come under strain as new voices are heard within the political arena, disturbing old corporatisms. And then there is the new environmental agenda with its demands for strategic planning action but with many uncertainties underpinning those demands. Planners do not have a clean sheet in responding to these demands. They have to maintain their professional identity and status and do so in a political and organizational context which has often been perceived as attacking that identity and status. Nevertheless there is clearly scope for change in professional practice, through the actions of the profession and, more importantly, other political actors either directly influencing practice within the planning department, within the local authority or on the broader planning policy front. For such change to be relevant in the twenty-first century, it should be built upon an understanding of the broader social and economic changes of the era, which we have outlined.

First, it is clear that the era of flexible accumulation will make new requirements of planners responding to or, more ambitiously, seeking to manage economic change in their locality. Since that economic change is occurring in a spatially diverse manner, this should involve much more finely grained knowledge on the planners' part about the economic characteristics of the locality and how they relate to broader spatial shifts in investment. While it is now accepted that such economic knowledge should be part of planners' training and practice, the suggestion here goes beyond general understanding of models of the location of economic activity or the subservience to market trends fostered by Thatcherism. Planning practice should further the study of local spatial economic processes and how markets, the state and space interact as an integral part of the local political process around urban change. It is not enough for planners alone to hold the knowledge of how economic processes are impacting on local spaces.

This is not to suggest that there should simply be a return to the 'report of survey' which preceded the structure plan, with its mass of information about local economic indicators. Rather, such knowledge should inform the much wider range of development planning, development control, project specific activities and general urban promotion that planning now encompasses, and such knowledge should be shared among a greater number of participants. The emphasis should be on understanding processes and local actors rather than collecting data: a critical edge is needed in engaging with local economic interests.

Rather than fostering a limited local corporatism, there is a need to consider alternative economic scenarios and the possibilities of achieving them. This clearly involves some commitment of resources but there is evidence of resources being made available for less well-conceived city promotion campaigns in the belief that they will encourage inward investment. Furthermore, the past experience of the Community Development Projects, and the work undertaken in many higher education institutions on local economies show that such understanding can be achieved even on tight budgets. The task is then to incorporate such understanding into everyday planning practice and the associated political processes.

The above rather assumes that economic development is a priority for planning and it is difficult to shift the consensus that local economic growth is an unqualified public good. But every pattern of economic development has distributive consequences and therefore the emphasis should be on producing alternative economic scenarios; each will have their own distributive pattern and a choice will need to be made between patterns. Such political debate could occur within the new political context which, as we have already argued, encourages the emergence of new voices and requires careful attention to protecting the position of minorities. This is a further possible task for planners, though not a task they can undertake on their own. Without national safeguards for minority voices and a more general sense of the limits to acceptable voices, any local action will be thrown on the defensive. But local state planners can play a limited role here and will need to overtly play this role if the liberal credentials that many of its adherents profess are to be maintained. This will involve an explicit attention to the needs of minorities, in the debate over alternative economic scenarios, the impact of specific projects, the choice of projected image for the locality, and the manner of the communication that planners engage in every day. These are all roles that planners within the local state can play.

There is also a need to expand the advocacy role of planners beyond the limited but important amount of planning aid that takes place and the highly paid advocacy that business interests can buy. Under increased conditions of polyphony, it would be a mistake to expect local government planners to be able to deal with all the voices and demands that arise, trading them off against each other in an objective manner. Rather, resources should be made available to provide a professional voice for community groups outside the local state so that they can engage in debate with others on more equal terms.

Finally, there is the question of the new environmental agenda and how planners should relate to it. There is the potential for the economic growth consensus to be undermined by certain voices from the 'deep

green' trend within environmentalism. The limitless pursuit of economic growth can be seen as inherently at odds with the maintenance
of environmental systems (Daly 1992). However, it is more likely that
the more technocratic tendencies within green thought will dominate the environmental agenda for planning (O'Riordan 1992). Here
the concern is how planners will engage with the varieties of expertise
about the environmental impacts of human activity. The danger appears
to be that planners, in pursuit of professional status, will themselves
seek to claim a particular expertise in the environmental area. This
produces the danger of ignoring the uncertainties about environmental
impacts and producing grand 'sustainable plans'.

In a postmodern era it would be more appropriate to recognize the
varieties of environmental expertise that exist, both within the scientific
communities and among the so-called 'lay' public (Dubash &
Oppenheimer 1992). The policy process then becomes a way of both
drawing on this loose network of expertise and activating it in many
small schemes and projects, rather than synthesizing it in a grand
scheme. The networks of the Environment City initiative, promoted
by the Royal Society for Nature Conservation, are very much in this
line. One of the strengths of this approach is the prominence it gives
to the voluntary sector and the innovative ideas that can arise from
the everyday experience of environmental impacts within this sector
(Rydin & Greig 1994); one of its weaknesses is in the implementation
of the ideas that arise.

In this context, the analytic approach we have adopted is particularly
apposite. Localized case study research provides evidence of the operation of planning in the specific circumstances of the locality but
the potential for endlessly diversified and varied local experiences is
constrained by the recognition that not everything is possible. Current
circumstances suggest a limited set of options with which to conduct
local experiments, hence the continued value of the concept of styles
of planning. The local experience of these styles itself can change the
nature of any one style, suggesting an evolution of new styles over
time. And the local experience is central to the way in which these
styles are invoked in the ongoing debate about planning. Repeated
evidence of the failure of, say, private sector management makes it
difficult to promote a remaking of planning further towards this style.
So in promoting our own ideas about the future direction of planning,
we would also call for more attention to those local cases where these
ideas are being experimented with and where the currently dominant
styles are being implemented. Planning is a local experience and both
research into the impacts and processes of planning, and political campaigning over new directions in planning have to look to the locality
for evidence, ideas and support for future change.

Bibliography

Adam Smith Institute 1983. *Omega Project: local government report*. London: Adam Smith Institute.

Alexander, A. 1982. *Local government in Britain since reorganisation*. London: Allen & Unwin.

Alty, R. & R. Darke 1987. A city centre for people: involving the community in planning for Sheffield's central area. *Planning Practice and Research* 3, September, 7–12.

Ambrose, P. 1986. *Whatever happened to planning?* London: Methuen.

Ambrose, P. & B. Colenutt 1975. *The property machine*. Harmondsworth: Penguin.

Anson, B. 1987. Covent Garden: an urban tragedy. *Architects' Journal* 22, July, 24–7.

Ash, M. 1977. *The crisis in planning*. London: Town & Country Planning Association.

Ball, M. 1983. *Housing policy and economic power*. London: Macmillan.

Barratt, S. & C. Fudge (eds) 1981. *Policy and action*. London: Methuen.

Best, R. 1981. *Land use and living space*. London: Methuen.

Blowers, A. 1986. Town planning – paradoxes and prospects. *The Planner* April, 11–18.

Boddy, M. 1984. Local economic and employment strategies. In *Local socialism*, M. Boddy & C. Fudge (eds). London: Macmillan.

Boddy, M., J. Lovering & K. Bassett 1986. *Sunbelt city? a study of economic change in Britain's M4 growth corridor*. Oxford: Clarendon Press.

Booth, S., D. Pitt & W. Money 1982. Organisational redundancy? A critical appraisal of the GEAR project. *Public Administration* 60, 56–72.

Boyle, R. (ed.) 1985. Leveraging urban development. *Policy and Politics* 13, 175–210.

Brindley, T. 1987. Community architecture – a triumph for Thatcherism? *Architectural Review* 181(1081), 4–6.

Brindley, T. & G. Stoker 1987. The politics of land-use planning on London's South Bank. In *Politics and policy-making in Britain*, L. Robins (ed.). London: Longman.

Broackes, N. 1984. The regeneration of London's Docklands. *Royal Society of Arts Journal* 132(5330), January, 105–17.

Broadbent, T. 1977. *Planning and profit in the urban economy*. London: Methuen.

Brook, C. A. 1983. *The rapidly changing field of high technology development and 'science parks'.* Paper presented to the School for Advanced Urban Studies/Oxford Polytechnic Land Policy Conference, March.

Cambridge and District Trades Council 1976. *Development in Cambridge.* Cambridge: Cambridge and District Trades Council.

Cambridge City Council 1966. *The future shape of Cambridge.* Cambridge: Cambridge City Council.

Cambridge Communist Party 1979. *Cambridge into the 1980s.* Cambridge: Cambridge Communist Party.

Cambridge University 1962. *The shape of Cambridge.* Cambridge: Cambridge University Press.

Campaign for Homes in Central London 1986. *City life: a future for Central London.* London: Campaign for Homes in Central London.

Carter, N. & C. Watts 1984. *The Cambridge Science Park.* London: Surveyors' Publications.

Christensen, T. 1979. *Neighbourhood survival.* London: Prism.

Clapham, D. & P. Kintrea forthcoming. Public housing. In *Regenerating the inner city,* D. Donnison (ed.). London: Routledge & Kegan Paul.

Coin Street Community Builders 1986. *The Coin Street development: briefing for council members.* London: Coin Street Community Builders.

Colchester Borough Council 1984. *Central area local plan.* Colchester: Colchester Borough Council.

Colenutt, B. & J. Lowe 1981. A London debate continued: does London need the Docklands Urban Development Corporation? *London Journal* 7(2), 235–8.

Colwell, R. 1984. Urban regeneration – how Scotland is showing the way. *120:Scottish Economic Development Review* 1 (4), 10.

Community Development Project 1977. *The costs of industrial change.* London: Home Office.

Coopers & Lybrand 1987. *Land-use planning and indicators of demand.* London: Coopers & Lybrand.

Cowan, R. 1986. The penny drops at Coin Street. *Roof* 11(2), March/April, 6–7.

Crummie, P. 1987. Colchester. *Current Archaeology* 103, 238–45.

Cullingworth, J. B. 1985. *Town and country planning in Britain,* 9th edn. London: Allen & Unwin.

Darlington, R. 1975. *Public participation in practice.* Cambridge: Cambridge City Council.

Davies, C. 1987. Ad hoc in the docks. *Architectural Review* 181(1080), February, 30–7.

Davies, J. G. 1972. *The evangelistic bureaucrat.* London: Tavistock.

DoE (Department of the Environment) 1975. *Review of the development control system: final report* (The Dobry report). London: HMSO.

DoE 1980. *Development – policy and practice.* Circular 22/80. London: HMSO.

DoE 1982. *Bringing in business.* London: HMSO.

DoE 1983. *Planning gain.* Circular 22/83. London: HMSO.

DoE 1984a. *Green belts.* Circular 14/84. London: HMSO.

BIBLIOGRAPHY

DoE 1984b. *Land for housing.* Circular 15/84.London: HMSO.

DoE 1984c. *Industrial development.* Circular 16/84. London: HMSO.

DoE 1985a. *The use of conditions in planning permissions.* Circular 1/85. London: HMSO.

DoE 1985b. *Lifting the burden.* Cmnd 9571. London: HMSO.

Doe 1986. *The future of development plans.* London: HMSO.

DoE 1987. *Housing – the government's proposals.* London: HMSO.

Docklands Consultative Committee 1985. *Four year review of the LDDC.* London: Docklands Consultative Committee.

Docklands Forum 1987. *Housing in Docklands.* London: Docklands Forum.

Docklands Joint Committee 1976. *London Docklands strategic plan.* London: Docklands Joint Committee.

Donnison, D. 1986. The lessons of GEAR. *Scottish Association of Geography Teachers Journal* 15, 18–22.

Dumbleton, B. 1976. *The second blitz: the demolition and rebuilding of town centres in South Wales.* Cardiff: Bob Dumbleton.

East Anglia Regional Strategy Team 1974. *Strategic choice for East Anglia.* East Anglia Regional Planning Board.

Elson, M. 1986. *Green belts.* London: Heinemann.

Essex County Council 1986. *Essex structure plan proposed alteration.* Chelmsford: Essex County Council.

Farnell, R. 1983. *Local planning in four English cities.* London: Gower.

Forrest, R., S. Lansley & A. Murie 1984. *A foot on the ladder?* WP 41. Bristol: School for Advanced Urban Studies.

Fothergill, S. & S. Gudgin 1982. *Unequal growth.* London: Heinemann.

Glasgow District Council 1984. *Barrowfield Project.* Glasgow: Glasgow District Council.

Glasgow District Council 1986. *GEAR local plan draft written statement.* Glasgow: Glasgow District Council.

Goldsmith, M. 1980. *Planning, politics and the city.* London: Hutchinson.

GLC (Greater London Council) 1976. *The future of the South Bank.* London: GLC.

GLC 1977. *Covent Garden action area plan.* London: GLC.

GLC 1981a. *Report of a meeting of the Planning and Communications Policy Committee.* 11 March, PC 772.

GLC 1981b. *The future of the South Bank wider area.* London: GLC.

GLC 1983. *Report of a meeting of the Joint Advisory Committee for Coin Street,* 11 October.

GLC 1984a. *London Docklands: review of the first two years operation of the LDDC.* London: GLC.

GLC 1984b. *The Greater London development plan – as proposed to be altered by the Greater London Council.* London: GLC.

GLC 1985a. *Community areas policy: a record of achievement.* London: GLC.

GLC 1985b. *The future of planning: London's proposals.* London: GLC.

GLC n.d. *What is popular planning?* London: GLC Popular Planning Unit.

Green, D. 1986. *The New Right: the counter revolution in political, economic and social thought.* Brighton: Wheatsheaf.

Grosskurth, A. 1984. Selling off the slums. *Roof* March/April, 25–7.

Hall, P. 1983. Enterprise zones and freeports. *New Society* 24 March, 460–2.
Hall, P., R. Thomas, H. Gracey, & R. Drewett 1973. *The containment of urban England*. London: Allen & Unwin.
Hardy, D. 1983a. *Making sense of the London Docklands: processes of change*. Geography and Planning Paper 9. Enfield: School of Geography and Planning, Middlesex Polytechnic.
Hardy, D. 1983b. *Making sense of the London Docklands: people and places*. Geography and Planning Paper 10. Enfield: School of Geography and Planning, Middlesex Polytechnic.
Hart, D. 1976. *Strategic planning in London: the rise and fall of the primary road network*. Oxford: Pergamon.
Haugh, P. 1986. US high technology multinationals and Silicon Glen. *Regional Studies* 20(2), 103–16.
Hawke, J. N. 1981. Planning agreements in practice. *Journal of Planning and Environment Law* January/February, 5–14, 86–97.
Healey, P. 1983. *Local plans in British land use planning*. Oxford: Pergamon.
Henney, A. 1985. *Trust the tenant*. Policy Study 68. London: Centre for Policy Studies.
Heseltine, M. 1983. *Reviving the inner cities*. London: Conservative Political Centre.
Heseltine, M. 1986. *Where there's a will*. London: Hutchinson.
Hillier, Parker, May & Rowden 1986. *Supplement to British shopping developments*. London: Hillier, Parker, May & Rowden.
Holford, W. & H. Myles-Wright 1950. *A report to Cambridge County Council*. Cambridge: Cambridge University Press.
House of Commons 1977. *Eighth report from the Expenditure Committee: planning procedures*. Volume 11: *Evidence*. London: HMSO.
House of Commons, Environment Committee 1984. *Green belt and land for housing*. London: HMSO.
House of Lords 1981. *Report from the Select Committee of the House of Lords on the LDDC. Area and constitution, London 1980*. London: HMSO.

Industry Department for Scotland 1987. *1986 Review of the Scottish Development Agency*. Edinburgh: Industry Department for Scotland.

Jenkin, P. 1984. Secretary of State's Address. *The Planner* 70(2), 15–17.
Jones, R. 1982. *Town and country chaos*. London: Adam Smith Institute.
Journal of Planning and Environment Law 1983. Notes of cases: GLC *et al.* v. Secretary of State for the Environment and Greycoat Commercial Estates Ltd. *Journal of Planning and Environment Law* December, 793–9.
Jowell, J. 1977. Bargaining in development control. *Journal of Planning and Environment Law* July, 414–33.
Joyce, P. 1986. Digging into a planning issue. *Planning* no. 681, 15 August, 10–11.

Keating, M. & R. Boyle 1986. *Remaking urban Scotland*. Edinburgh: Edinburgh University Press.
Kirk, G. 1980. *Urban planning in a capitalist society*. London: Croom Helm.

Knight, Frank & Rutley 1987. *Docklands: commercial and residential develop-ments*. London: KFR Research.

Knowsley Metropolitan Borough Council 1986. Note summarizing *The Stockbridge Village Investigation: and evaluation of the costs, benefits and lessons*. Roger Tym & Partners, 9 April.

Lambert, J., C. Paris & B. Blackaby 1978. *Housing policy and the state*. London: Macmillan.

Lawless, P. 1981. *Britain's inner cities*. London: Harper & Row.

Leach, S, & N. Moore 1979. County–district relations in shire and metro-politan counties in the field of town and country planning: a comparison. *Policy & Politics* **7**, 271–9.

Lebas, E. 1983. The state in British and French urban research or the crisis of the urban question. In *Urban social research: problems and prospects*, V. Pons & R. Francis (eds). London: Routledge & Kegan Paul.

Leclerc, R. & D. Draffan 1984. The Glasgow Eastern Area Renewal project. *Town Planning Review* **55**, 335–51.

Ledgerwood, G. 1985. *Urban innovation*. London: Gower.

Lloyd, M. 1984. Enterprise zones: the evaluation of an experiment. *The Planner* **70**, 23–5.

London Borough of Lambeth 1975. *Waterloo draft planning strategy*. London: London Borough of Lambeth.

London Borough of Lambeth 1977. *Waterloo district plan*. London: London Borough of Lambeth.

London Borough of Newham 1986a. *Beckton local plan*. London: London Borough of Newham.

London Borough of Newham 1986b. *South Docklands draft local plan*. London: London Borough of Newham.

London Borough of Southwark 1985. *The North Southwark draft local plan*. London: London Borough of Southwark.

LDDC (London Docklands Development Corporation) 1986a. *Corporate plan – April 1986*. London: LDDC.

LDDC 1986b. *Annual report and accounts 1985/86*. London: LDDC.

LDDC 1987. *Annual review 1986/87*. London: LDDC.

London Docklands Study Team 1973. *Docklands redevelopment proposals for East London*. London: R. Travers Morgan & Partners.

Loughlin, M. 1980. The scope and importance of 'material considerations'. *Urban Law and Policy* **3**(2), June, 171–92.

Lowe, P. 1986. *Urban social movements: the city after Castells*. London: Macmillan.

Martin, R. 1986. In what sense a 'jobs boom': employment recovery, government policy and the regions. *Regional Studies* **20**(5), 466–7.

Massey, D. 1982. Enterprise zones: a political issue. *International Journal of Urban and Regional Research* **6**, 429–34.

Massey, D. 1984. *Spatial divisions of labour*. London: Macmillan.

McDonald, A. 1986. *The Weller way*. London: Faber & Faber.

Middleton, A. 1985. The growth of the east end of Glasgow. *The Planner* **71**(2), 8–9.

Milne, R. 1979. The battle of Coin Street. *Estates Gazette* no. 252, 22/29 December, 1167–9.

Milne, R. 1984. Lords plea as Coin Street scheme struggles. *Planning* no. 550, 6 January, 6.

Ministry of Housing & Local Government 1969. *People and planning* (the Skeffington report). London: HMSO.

Moor, N. 1984. A framework for change. *The Planner* 70(5), 15–19.

Morton, J. 1984. Transformation: how an unpopular estate is being transformed by Stockbridge Village Trust. *Local Government News* September.

Mott Committee 1969. Report of the sub-committee on the relationship between the University and science-based industry. *Cambridge University Reporter* 22 October, 370–6.

Munton, R. 1983. *London's Green Belts: containment in practice*. London: Allen & Unwin.

Nelson, R. 1980. *Participating in GEAR*. Glasgow: University of Strathclyde.

Newham Docklands Forum 1983. *The people's plan for the Royal Docks*. London: Newham Docklands Forum.

Newman, I. & M. Mayo 1981. Docklands. *International Journal of Urban and Regional Research* 5 (4), 529–45.

Nuffield Commission of Inquiry 1986. *Town and country planning* (the Flowers report). Oxford: Nuffield Foundation.

Pahl, R. E. 1975. *Whose city?* 2nd edn. Harmondsworth: Penguin.

Parry Lewis, J. 1974. *The Cambridge sub-region*. Cambridge: Cambridge County Council.

Parsons, D. 1986. *The political economy of British regional policy*. London: Croom Helm.

Pawley, M. 1986. Electric city of our dreams. *New Society* 13 June, 12–14.

Pickvance, C. 1981. Physical planning and market forces in urban development. In *Critical readings in planning theory*, C. Paris (ed.). Oxford: Pergamon.

Potter, S. 1986. New town statistics. *Town and Country Planning* 55(11), 304–9.

Power, A. 1987. *The priority estates project model*. London: DoE.

Property Advisory Group 1980. *The structure and activity of the property development industry*. London: DoE.

Ravetz, A. 1980. *Remaking cities: contradictions of the recent urban environment*. London: Croom Helm.

Ravetz, A. 1986. *The government of space: town planning in modern society*. London: Faber & Faber.

Reade, E. 1982. Section 52 and corporatism in planning. *Journal of Planning and Environment Law* January, 8–16.

Reade, E. 1987. *British town and country planning*. Milton Keynes: Open University Press.

Rees, G. &. J. Lambert 1985. *Cities in crisis: the political economy of urban development*. London: Edward Arnold.

Richard Rogers & Partners 1981. Coin Street development. *Architectural Design* no. 3/4, 52–5.

BIBLIOGRAPHY

Ridley, N. 1986. Letter to the Chairman of SERPLAN. In SERPLAN (1986).

Robinson, J. R. & M. G. Lloyd 1986. Lifting the burden: a means or an end? *Local Government Studies* 12(3), 51–64.

Rotherhithe Community Planning Centre 1986. *Our Side of the River: community views on the redevelopment of Southwark's dockland.* London: Rotherhithe Community Planning Centre.

Rydin, Y. 1986. *Housing land policy.* London: Gower.

Rydin, Y. 1988. Joint housing studies: housebuilders, planners and the availability of land. *Local Government Studies* 14(2), 69–80.

SDA (Scottish Development Agency) 1978. *The future of GEAR.* Glasgow: SDA.

SDA 1979. *GEAR overall proposals.* Glasgow: SDA.

SDA 1980. *GEAR strategy and programme.* Glasgow: SDA.

SDA 1986. *GEAR project progress statement.* Glasgow: SDA.

SDA 1987. *Glasgow's East End: a continuing commitment.* Glasgow SDA.

Segal, Quince & Partners 1985. *The Cambridge Phenomenon.* London: Segal, Quince & Partners.

Self, P. 1979. The siege of Coin Street. *Town and Country Planning* August, 149–50.

Senior, D. 1956. *A guide to the Cambridge plan.* Cambridge:Cambridge County Council.

SERPLAN 1985. *Developing South East regional strategy: South East England in the 1990s: a regional statement.* RPC 450. London: SERPLAN.

SERPLAN 1986. *Regional strategic guidance.* RPC 602. London: SERPLAN.

Sheffield City Council 1987. *Sheffield working it out.* Sheffield: Sheffield City Council.

Simmie, J. 1974. *Citizens in conflict.* London: Hutchinson.

Simmie, J. 1981. *Power, property and corporatism.* London: Macmillan.

Simmie, J. 1985. Corporatism and planning. In *The political economy of corporatism,* W. Grant (ed.). London: Macmillan.

Sorensen, A. 1983. Planning comes of age: a libertarian perspective. *The Planner* 68, 184–8.

South East Regional Planning Council 1964. *South East study.* London: HMSO.

South East Regional Planning Council 1967. *Strategic plan for the South East.* London: HMSO.

Spencer, K., A. Taylor, B. Smith, J. Mawsin, N. Flynn & R. Battey 1986. *Crisis in the industrial heartland: a study of the West Midlands.* Oxford: Clarendon Press.

Stoker, G. 1985. The building societies and the Conservatives' housing strategy into the late 1980s. *Critical Social Policy* 12, Spring, 63–8.

Stoker, G. 1988. *The politics of local government.* London: Macmillan.

Stretton, H. 1978. *Urban planning in rich and poor countries.* Oxford: OPUS.

Sudjic, D. & S. Wood 1981. Last chance for the Thames. *Sunday Times Magazine* 19 April, 21–31.

Thornley, A, 1981. *Thatcherism and town planning.* Planning Studies No. 12. London: Polytechnic of Central London.

Thornley, A. 1986. *Thatcherism and simplified planning regimes*. Paper presented to the Research in Development Planning seminar, School for Advanced Urban Studies, Bristol, June.

Tym, R. & Partners 1983. *Monitoring enterprise zones: year 2 report*. London: Roger Tym & Partners.

Underwood, J. 1980. *Town planners in search of a role*. OP 6. Bristol: School for Advanced Urban Studies.

Underwood, J. 1981. Development control: a case study of discretion in action. In *Policy and action*, S. Barrett & C. Fudge (eds). London: Methuen.

Wainright, H. 1985. Sharing power: popular planning and the GLC. In *Going Local? 2*. London: Polytechnic of Central London.

Wainright, H. 1987. *Labour: a tale of two parties*. London: Hogarth.

Ward, R. 1986. London: the emerging docklands city. *Built Environment* 12(3), 117–27.

Wates, N. 1976. *The battle for Tolmers Square*. London: Routledge.

Wates, N. (ed.) 1986. *The Limehouse petition*. London: Limehouse Development Group.

Wates, N. & C. Knevitt 1987. *Community architecture*. London: Penguin.

Young, S. 1985. *The nature of privatisation*. Annual conference of the Political Studies Association, University of Manchester, 16–18 April.

Young, S. 1986. The nature of privatisation in Britain, 1979–85. *West European Politics* 9(2), 235–52.

Additions to bibliography in the second edition

Agyeman, J. & B. Evans 1994. *Local Environmental Policies and Strategies*. London: Longman.

Ambrose, P. 1994. *Urban Process and Power*. London: Routledge.

Ashworth, G. & H. Voogd 1990. *Selling the city: marketing approaches in public sector urban planning*. London: Bellhaven.

Bailey, N., A. Barker & K. McDonald 1995. *Partnership Agencies in British Urban Policy*. London: UCL Press.

Ball, R. & A. Pratt 1994. *Industrial Property: policy and economic development*. London: Routledge.

Barnes, J. 1990. Urban Development Corporations: the lessons from Docklands. In *Radical Planning Initiatives*. J. Montgomery and A. Thornley (eds), Aldershot: Gower.

Beck, U., A. Giddens & S. Lash 1994. *Reflexive Modernisation*. Cambridge: Polity Press.

Bianchini, F. & M. Parkinson 1993. *Cultural Policy and Urban Regeneration*. Manchester: Manchester University Press.

Boddy, M. & C. Fudge 1984. *Local Socialism? Labour Councils and New Left Alternatives*. London: Macmillan.

Brindley, T. & G. Stoker 1988. Partnership in inner city urban renewal – a critical analysis. *Local Government Policy Making* **15** (2), September, 3–12.

Brownill, S. 1990. *Developing London's Docklands: another great planning disaster?* London: Paul Chapman.

Commission for Local Democracy 1995. *Taking Charge: the rebirth of local democracy*. London: CLD.

Coupland, A. 1992. Docklands: dream or disaster? In *The Crisis of London*. A. Thornley (ed.). London: Routledge.

Daly, H. 1992. *Steady-State Economics*. London: Earthscan.

Dobson, A. (ed.) 1991. *The Green Reader*. London: André Deutsch.

Docklands Forum 1993. *Race and Housing in London's Docklands*. London: Docklands Forum.

Dubash, N. & M. Oppenheimer 1992. Modifying the mandate of existing institutions: NGOs. In *Confronting Climate Change: risks, implications and responses*. I. Mintzer (ed.). Cambridge: Cambridge University Press.

Dunsire, A. 1993. Modes of governance. In *Modern Governance*. J. Kooiman (ed.). London: Sage.

Eckersley, R. 1992. *Environmentalism and Political Theory*. London: UCL Press.

Edwards, B. 1992. *London Docklands: urban design in an age of deregulation*. Oxford: Butterworth-Heinemann.

Evans, B. 1993. Why we no longer need a town planning profession. *Planning Practice & Research* **8**, 9–15.

Evans, R. & H. Russell 1993. *Stockbridge Village: achievements and lessons after 10 years*. Liverpool: European Institute for Urban Affairs, Liverpool John Moores University.

Fainstein, S. 1994. *The City Builders*. Oxford: Blackwell.

Giddens, A. 1990. *The Consequences of Modernity*. Cambridge: Polity Press.

BIBLIOGRAPHY

Gyford, J. 1995. The politics and planning of London. In *Planning London*. J. Simmie (ed.). London: UCL Press.

Hansen, A. (ed.) 1993. *The Mass Media and Environmental Issues*. Leicester: LUP.

Harvey, D. 1989. *The Condition of Postmodernity*. Oxford: Blackwell.

Healey, P. et al. (eds) 1992. *Rebuilding the City: property-led urban regeneration*. London: E & F Spon.

Healey, P. & T. Shaw 1994. Changing meanings of the 'environment' in the British planning system. *Transactions of the Institute of British Geographers* **19** (4), 425–38.

Heskin, A. D. 1992. Ethnicity, race, class and ideology come together in LA. In *After Modernism: global restructuring and the changing boundaries of city life*. M. Smith (ed.). New Jersey: Transaction Publishers.

Independent 1995. Docklands: a special report. 25 January.

Jacobs, J. 1961. *The Death and Life of Great American Cities*. New York: Random House.

Jessop, B. 1995. The regulation approach, governance and post-Fordism, *Economy & Society*, **24** (3), August, 307–33.

Krier, L. et al. 1978. *Rational Architecture*. Brussels: Editions Archives d'Architecture Moderne.

Lash, S. & J. Urry 1987. *The End of Organised Capitalism*. Cambridge: Polity Press.

LDDC (London Docklands Development Corporation) 1995. *LDDC, Key Facts and Figures*. London: LDDC.

Massey, D. 1984. *Spatial Divisions of Labour: social structure and the geography of production*. London: Macmillan.

Massey, D. & J. Allen 1988. *The Economy in Question*. London: Sage.

Massey, D., P. Quintas & D. Wield 1992. *High Tech Fantasies*. London: Routledge.

Meegan, R. 1988, A crisis of mass production? In D. Massey & J. Allen *op. cit.*

Myerson, G. & Y. Rydin 1994. Environment and Planning: a tale of the mundane and the sublime, *Society & Space* **12**, 437–52.

O'Riordan, T. 1992. The environment. In *Policy and Change in Thatcher's Britain*. P. Cloke (ed.). Oxford: Pergamon.

Rhodes, R. 1995. The new governance: governing without government. In *The State of Britain Seminars (11)*. Swindon: ESRC.

Rydin, Y. 1995. Sustainable development and the role of land use planning. *Area* **27** (4), 369–77.

Rydin, Y. & A. Greig 1994. Talking past each other: environmentalists in different organisational settings. *Environmental Politics* **4** (2), 271–94.

Scoffham, E. R. 1982. *The Shape of British Housing*. London: George Godwin.

Zukin, S. 1982. *Loft Living: culture and capital in urban change*. Baltimore: Johns Hopkins University Press.

Index

225

INDEX

INDEX

23–5, 139–57, 158, 162, 166, 175, 178, 196

Private Sector Villages 39, 41–2

Privatisation 24, 155

Property Advisory Group 11, 12

Public Investment Planning 9, 11–13, 22–3, 121–38, 158, 160, 165, 173–4, 179, 196

Public Participation 4, 17–18, 27, 33–4, 89, 136, 154, 161, 166, 168

Regional Planning 32, 43, 53, 65, 96

Regulative Planning 9, 11, 13–16, 27–50, 158–60, 164, 168–9, 177, 179

Responsive Planning 197–200

Scottish Development Agency 22, 121, 123, 124–6, 127, 130, 131, 132, 133–5, 160, 163

Scottish Special Housing Association 123–4, 125–7

Section 52 Agreements 35, 37

Simplified Planning Zones 1, 16–17, 51, 178

South Bank Employers' Group 207

South Bank Forum 208

South Cambridgeshire District Council 33, 34, 35, 37, 38–9, 43, 46–7

Sports Council 130

Stockbridge Village Housing Association 144, 150–2, 153–4

Stockbridge Village Trust 24, 139–57, 160, 161–2, 175, 178, 182, 196

Strathclyde Regional Council 123, 125, 127, 130, 132

Thamesmead 24, 156

The Thatcher Government 1, 2, 16, 20, 26, 158, 163, 170, 176, 182

Trend Planning 9, 11, 15–17, 51–73, 158–9, 165, 169–70, 182

Urban Development Corporations 1, 2, 21, 96–120, 161, 178, 182

Urban Development Grant 20–1

Urban Gatekeepers 68, 158

Use Classes Order 37, 178

Waterloo Community Development Group 77–8, 80, 90, 208